CIHM
Microfiche
Series
(Monographs)

ICMH
Collection de
microfiches
(monographies)

Canadian Institute for Historical Microreproductions / Institut canadien de microreproductions historiques

©1997

Technical and Bibliographic Notes / Notes techniques et bibliographiques

The Institute has attempted to obtain the best original copy available for filming. Features of this copy which may be bibliographically unique, which may alter any of the images in the reproduction, or which may significantly change the usual method of filming are checked below.

- [x] Coloured covers / Couverture de couleur
- [ ] Covers damaged / Couverture endommagée
- [ ] Covers restored and/or laminated / Couverture restaurée et/ou pelliculée
- [ ] Cover title missing / Le titre de couverture manque
- [ ] Coloured maps / Cartes géographiques en couleur
- [ ] Coloured ink (i.e. other than blue or black) / Encre de couleur (i.e. autre que bleue ou noire)
- [ ] Coloured plates and/or illustrations / Planches et/ou illustrations en couleur
- [ ] Bound with other material / Relié avec d'autres documents
- [ ] Only edition available / Seule édition disponible
- [ ] Tight binding may cause shadows or distortion along interior margin / La reliure serrée peut causer de l'ombre ou de la distorsion le long de la marge intérieure.
- [ ] Blank leaves added during restorations may appear within the text. Whenever possible, these have been omitted from filming / Il se peut que certaines pages blanches ajoutées lors d'une restauration apparaissent dans le texte, mais, lorsque cela était possible, ces pages n'ont pas été filmées.
- [x] Additional comments / Commentaires supplémentaires: Various pagings.

L'Institut a microfilmé le meilleur exemplaire qu'il lui a été possible de se procurer. Les détails de cet exemplaire qui sont peut-être uniques du point de vue bibliographique, qui peuvent modifier une image reproduite, ou qui peuvent exiger une modification dans la méthode normale de filmage sont indiqués ci-dessous.

- [ ] Coloured pages / Pages de couleur
- [ ] Pages damaged / Pages endommagées
- [ ] Pages restored and/or laminated / Pages restaurées et/ou pelliculées
- [x] Pages discoloured, stained or foxed / Pages décolorées, tachetées ou piquées
- [ ] Pages detached / Pages détachées
- [x] Showthrough / Transparence
- [ ] Quality of print varies / Qualité inégale de l'impression
- [ ] Includes supplementary material / Comprend du matériel supplémentaire
- [ ] Pages wholly or partially obscured by errata slips, tissues, etc., have been refilmed to ensure the best possible image / Les pages totalement ou partiellement obscurcies par un feuillet d'errata, une pelure, etc., ont été filmées à nouveau de façon à obtenir la meilleure image possible.
- [ ] Opposing pages with varying colouration or discolourations are filmed twice to ensure the best possible image / Les pages s'opposant ayant des colorations variables ou des décolorations sont filmées deux fois afin d'obtenir la meilleure image possible.

This item is filmed at the reduction ratio checked below / Ce document est filmé au taux de réduction indiqué ci-dessous.

10x	14x	18x	22x	26x	30x
12x	16x	20x	24x	28x	32x

The copy filmed here has been reproduced thanks to the generosity of:

    National Library of Canada

The images appearing here are the best quality possible considering the condition and legibility of the original copy and in keeping with the filming contract specifications.

Original copies in printed paper covers are filmed beginning with the front cover and ending on the last page with a printed or illustrated impression, or the back cover when appropriate. All other original copies are filmed beginning on the first page with a printed or illustrated impression, and ending on the last page with a printed or illustrated impression.

The last recorded frame on each microfiche shall contain the symbol —▶ (meaning "CONTINUED"), or the symbol ▽ (meaning "END"), whichever applies.

Maps, plates, charts, etc., may be filmed at different reduction ratios. Those too large to be entirely included in one exposure are filmed beginning in the upper left hand corner, left to right and top to bottom, as many frames as required. The following diagrams illustrate the method:

L'exemplaire filmé fut reproduit grâce à la générosité de:

    Bibliothèque nationale du Canada

Les images suivantes ont été reproduites avec le plus grand soin, compte tenu de la condition et de la netteté de l'exemplaire filmé, et en conformité avec les conditions du contrat de filmage.

Les exemplaires originaux dont la couverture en papier est imprimée sont filmés en commençant par le premier plat et en terminant soit par la dernière page qui comporte une empreinte d'impression ou d'illustration, soit par le second plat, selon le cas. Tous les autres exemplaires originaux sont filmés en commençant par la première page qui comporte une empreinte d'impression ou d'illustration et en terminant par la dernière page qui comporte une telle empreinte.

Un des symboles suivants apparaîtra sur la dernière image de chaque microfiche, selon le cas: le symbole —▶ signifie "A SUIVRE", le symbole ▽ signifie "FIN".

Les cartes, planches, tableaux, etc., peuvent être filmés à des taux de réduction différents. Lorsque le document est trop grand pour être reproduit en un seul cliché, il est filmé à partir de l'angle supérieur gauche, de gauche à droite, et de haut en bas, en prenant le nombre d'images nécessaire. Les diagrammes suivants illustrent la méthode.

| 1 | 2 | 3 |

| 1 |
| 2 |
| 3 |

| 1 | 2 | 3 |
| 4 | 5 | 6 |

MICROCOPY RESOLUTION TEST CHART

(ANSI and ISO TEST CHART No. 2)

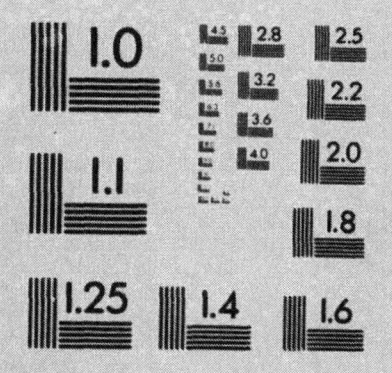

APPLIED IMAGE Inc
1653 East Main Street
Rochester, New York 14609 USA
(716) 482 - 0300 - Phone
(716) 288 - 5989 - Fax

CANADA
DEPARTMENT OF MINES
Hon. P. E. Blondin, Minister; R. G. McConnell, Deputy Minister.

MINES BRANCH
Eugene Haanel, Ph.D., Director.

RESEARCHES ON COBALT AND COBALT ALLOYS, CONDUCTED AT QUEENS UNIVERSITY, KINGSTON, ONTARIO, FOR THE MINES BRANCH OF THE DEPARTMENT OF MINES

PART IV.

Cobalt Alloys with Non-Corrosive Properties

BY

Herbert T. Kalmus, B.Sc., Ph.D.
and
K. B. Blake, B.Sc.

NATIONAL RESEARCH COUNCIL

OTTAWA
Government Printing Bureau
1916

No. 411

CANADA
DEPARTMENT OF MINES
Hon. P. E. Blondin, Minister; R. G. McConnell, Deputy Minister.

MINES BRANCH
Eugene Haanel, Ph.D., Director.

RESEARCHES ON COBALT AND COBALT ALLOYS, CONDUCTED AT QUEENS UNIVERSITY, KINGSTON, ONTARIO, FOR THE MINES BRANCH OF THE DEPARTMENT OF MINES

PART IV.

Cobalt Alloys with Non-Corrosive Properties

BY

Herbert T. Kalmus, B.Sc., Ph.D.
and
K. B. Blake, B.Sc.

OTTAWA
Government Printing Bureau
1916

No. 411

LETTER OF TRANSMITTAL

Dr. Eugene Haanel,
    Director Mines Branch,
        Department of Mines,
            Ottawa, Canada.

Sir,—
    I beg to submit, herewith, a report on "Cobalt Alloys with Non-Corrosive Properties." This is the fourth completed part of the series of investigations on cobalt and cobalt alloys for the purpose of increasing their economic importance, which has been the subject of the special researches conducted under my direction at Queens University, Kingston, Ontario, for the Mines Branch of the Department of Mines, Ottawa.

        I have the honour to be, Sir,
            Your obedient servant,
                (*Signed*) **Herbert T. Kalmus.**

Boston, Mass., April 22, 1916.

CONTENTS

	PAGE
Introductory	1
Corrosion in general	1
Purpose of the investigation	3
Preliminary experiments	3
Conclusions	4
Experiments on non-corrosive alloys prepared by additions of cobalt, nickel, and copper, to American ingot iron	4
Materials for preparation of alloys	4
Method of preparation of alloys	5
Crucible and furnace	5
Melting and casting	5
Preparation of discs for corrosion tests	6
Method of measurement to determine corrosion	6
Absorption of carbon monoxide gas	6
Corrosion tests—	
Series I	7
Data of corrosion experiments	7
Table of particulars of corrosion tests	9
Conclusions	11
Series II	12
Data of corrosion experiments	12
Table of particulars of corrosion tests	12
Conclusions	15
Series III	16
Preparation of alloys for furnace treatment	16
Data of corrosion experiments	16
Table of particulars of corrosion tests—	
Kingston samples	24
American Rolling Mill samples	29
Microphotographs: Plates III–XXXI	30
Series IV	30
Accelerated corrosion tests	30
Method of making accelerated corrosion tests	30
Accelerated corrosion measurements	31
Conclusion	33
Corrosion tests, American ingot iron alloys in the form of sheet roofing materials	34
Analyses	34
Graphical representation of data: Diagrams i–vii	34
General conclusions	35
Acknowledgments	36
Catalogue of Mines Branch publications	End

ILLUSTRATIONS.

Photographs.

Plate			PAGE
I.	Corrosion tests: method of exposing cobalt alloy discs		6
II.	" " " " " " "		6
III.	Microphotograph: alloy B 202		30
IV.	" " B₂ 199		30
V.	" " S 250		30
VI.	" " S 262		30
VII.	" " S 255		30
VIII.	" " B 209		30
IX.	" " S 252		30
X.	" " S 260		30
XI.	" " B 198		30
XII.	" " B₂ 200		30
XIII.	" " C 202		30
XIV.	" " C 255		30
XV.	" " B 199		30
XVI.	" " S 251		30
XVII.	" " S 254		30
XVIII.	" " S 258		30
XIX.	" " S 259		30
XX.	" " B₂ 195		30
XXI.	" " S 263		30
XXII.	" " S 269		30
XXIII.	" " B 195		30
XXIV.	" " B 196		30
XXV.	" " B₂ 197		30
XXVI.	" " B 197		30
XXVII.	" " C 204		30
XXVIII.	" " B₂ 205		30
XXIX.	" " B₂ 206		30
XXX.	" " B 206		30
XXXI.	" " B 208		30

Drawings.

CORROSION TESTS—
Diagram i.
 Series I. Different percentages of same materials.
  Figure 1. Copper.. 34
   " 2. Cobalt... 34
   " 3. Pure iron.. 34
 Series II. Different percentages of same materials.
  Figure 3. Pure iron.. 34
   " 4. Nickel... 34

Diagram ii.
 Series III. Different percentages of same materials.
  Figure 5. Copper: low carbon.................................. 34
   " 6. Nickel: " "... 34
   " 7. Cobalt: " "... 34
   " 8. Copper: medium carbon............................ 34
   " 9. Nickel: " "... 34
   " 10. Cobalt: " "... 34
   " 11. Nickel: high carbon................................. 34
   " 12. Cobalt: " "... 34

Diagram iii.
 Series III. Different percentages of carbon.
  Figure 13. Cobalt 0·25%.. 34
   " 14. " 0·35%.. 34
   " 15. " 0·50%.. 34
   " 16. Nickel 0·25%.. 34
   " 17. " 0·35%.. 34
   " 18. " 0·50%.. 34
   " 19. Copper 0·25%... 34
   " 20. " 0·35%.. 34
   " 21. " 0·50%.. 34

vii

Diagram iv.
   Series III. Different percentages of carbon.   PAGE
      Figure 22. Cobalt 0·75%.. 34
      " 23. " 1·0 %.. 34
      " 24. Nickel 1·0 %.. 34
      " 25. " 0·75%.. 34
      " 26. Cobalt 2·0 %.. 34
      " 27. Nickel 2·0 %.. 34
      " 28. Copper 0·75%.. 34
      " 29. Cobalt 3·0 %.. 34
      " 30. Nickel 3·0 %.. 34

Diagram v.
   Series III. Same percentages of different materials.
      Figure 31. Alloys 0·25% carbon low 0·0 – 0·17 34
      " 32. " 0·35% " " 34
      " 33. " 0·50% " " 34
      " 34. " 0·25% carbon medium 0·18 – 0·25 34
      " 35. " 0·35% " " 34
      " 36. " 0·50% " " 34
      " 37. " 0·25% carbon high 0·26 – 34
      " 38. " 0·35% " " 34
      " 39. " 0·50% " " 34

Diagram vi.
   Series III. Same percentages of different materials.
      Figure 40. Alloys 0·75% carbon low........................... 34
      " 41. " 1·0 % " " 34
      " 42. " 2·0 % " " 34
      " 43. " 0·75% carbon medium 34
      " 44. " 1·0 % " " 34
      " 45. " 2·0 % " " 34
      " 46. " 1·0 % carbon high....................... 34
      " 47. " 2·0 % " " 34

Diagram vii.
   Series III. Same percentages of different materials.
      Figure 48. Alloys 3·0 % carbon low........................... 34
      " 49. " 3·0 % carbon medium....................... 34
      " 50. " 3·0 % carbon high........................... 34

RESEARCHES ON COBALT AND COBALT ALLOYS, CONDUCTED AT QUEENS UNIVERSITY, KINGSTON, ONTARIO, FOR THE MINES BRANCH OF THE DEPARTMENT OF MINES

PART IV

COBALT ALLOYS WITH NON-CORROSIVE PROPERTIES

PART IV

COBALT ALLOYS WITH NON-CORROSIVE PROPERTIES

INTRODUCTORY

This investigation is part IV of the series of researches which have been conducted at Queens University, Kingston, Ontario, for the Mines Branch, of the Department of Mines, Ottawa, with a view to finding increased commercial uses for cobalt and its alloys. The investigations have been undertaken under the following subdivisions:—

I. Preparation of Metallic Cobalt by Reduction of the Oxide.
II. The Physical Properties of the Metal Cobalt.
III. Electro-Plating with Cobalt.
IV. Cobalt Alloys with Non-Corrosive Properties.
V. Magnetic Properties of Cobalt and of Fe_2Co.

CORROSION IN GENERAL

Of the three principal theories of corrosion: (1) the carbonic acid theory; (2) the hydrogen peroxide theory; and (3) the electrolytic theory, the latter seems to fit all the facts most completely.

W. H. Walker[1] gives the essential facts of the electrolytic explanation of corrosion as follows:—

Every metal when placed in water, or under such conditions that a film of water may condense upon it, tends to dissolve in the water, or, in other words, to pass from its atomic or metallic condition into its ionic condition. This escaping tendency of the metals varies from that shown by sodium or potassium, which is so great as to cause instant and rapid decomposition of the metal and water, to gold or platinum where such tendency to dissolve is zero. Between these two extremes we find the other common metals, including thereunder the element hydrogen, which may be considered as metal. As the atom of metal passes into the water, it assumes a positive charge of electricity, leaving the metallic mass from which it separated charged negatively; this property or escaping tendency of the metal is termed its solution pressure. It is obvious, however, that this action can continue for only a short time; owing to the fact that the mass of metal and solution are of opposite polarity, the electrolytic tension becomes so great that no more atoms can escape to the ionic state, and the solvent action ceases. This condition was first described by Helmholtz, and called by him an electrolytic double layer. If now there be in the water ions of another metal which has a smaller solution pressure than the one under consideration, the action as above described will be reversed and the ion with the less solution pressure will pass back to the metallic state, plating out on the first metal and giving up its charge of electricity. At this point the first metal will be charged positively, and the solution in the immediate vicinity negatively, and there will tend to be set up a second electrolytic double layer opposite in polarity to the first. The result is, a current of electricity flows from the metal to the solution at the point where the metal passes into solution, through the solution to the metal at the point where the ions of the second metal are plating out, and back through the first metal to the starting point again. The electrolytic double layers are thus destroyed, an electric current passes, and the solvent action of the water on the first metal continues.

[1] Journal of the Iron and Steel Institute, 1909, Vol. I, p. 70.

This phenomenon and its relation to the corrosion of iron are clearly exemplified in the well known Daniel or gravity cell. In the case of pure iron in water a perfectly analogous condition is found to exist. Water itself is dissociated to a small but perfectly definite extent into its ions, hydrogen (H) and hydroxyl (OH). When a strip of pure iron comes into contact with water, it sends into the water iron atoms in the form of positively charged ions. Hydrogen as a metal has a much smaller solution pressure than iron, and hence an equivalent number of hydrogen ions plate out on the iron strip (leaving the free hydroxyl ions with their negative charges to balance the iron ions with their positive charges), and an electric current flows from the iron by means of the iron ions to the solution, and by means of the hydrogens from the solution back to the iron again, thus completing the circuit. But here comes an important break in the analogy of the film of copper in the Daniel cell. Deposited copper is a good conductor of the current, and offers no resistance to its flow from the solution to the iron on which it is attached. The reverse is true of the deposited hydrogen; here we have a high insulator—a film of gas which offers a great resistance to the flow of the current. Hence, although in the case of the iron strip in water all the conditions for continuous solution are present, owing to the resistance offered by the deposited hydrogen film (called polarization) the action must cease.

Just as in the case of iron in a copper sulphate solution, the rapidity of the action depended upon the number of copper ions present in the solution, so here the solution of the iron, in the first instance, depends upon the number of the hydrogen ions present. This number of hydrogen ions, or the concentration of these ions, is increased by the addition of any acid. So weak an acid as carbonic increases the number, but to a relatively small amount; while a strong acid, like hydrochloric or sulphuric, adds to the number to such an extent that the solvent action becomes violent, and the deposited hydrogen comes off as a stream of gas.

Since the presence of the polarizing film of hydrogen arrests the further solution of the iron, it is obvious that in order for the reaction to proceed this hydrogen must be removed. The destruction of the hydrogen film in ordinary corrosion is accomplished by the oxygen of the atmosphere, which is dissolved in the water. The action here taking place is a simple union of the hydrogen on the iron and the oxygen dissolved from the air, with the re-formation of water. It follows from this that any substance which dissolves or reacts with hydrogen should accelerate corrosion. This is found to be in fact the case.

The trend of scientific and engineering opinion at present is, to accept this electrolytic explanation of corrosion. The truth concerning corrosion is, no doubt, much more complex than the above statement of electrolysis. As a theory it has been disappointing to engineers, for it has failed to predict many disquieting corrosion phenomena, or to suggest means of remedy. Nevertheless, for the present, it is the best working hypothesis that we have.

Accepting the electrolytic theory, it follows in the case of iron—other conditions being alike—that the approach to absolute freedom from impurities should add to its resistance to corrosion. It does not follow, however, that of the metals technically produced, those showing, under analysis, the least amount of impurities, are the most non-corrosive, for other conditions are by no means always alike. Even in a pure metal, stresses or strains produced by uneven cooling of a casting, or by rolling, whether hot or cold, provide unequal solution tension at various points in the metal, particularly on the surface, and would thus promote corrosion. Also in the preparation of metals of high chemical purity, the resistance to corrosion may be decreased by gas occlusion, or in other ways, in part due to this very attempt to attain a high degree of purity.

Stresses and strains are, usually, partially or entirely overcome by thoroughly annealing metals to be used for sheet roofing materials, or for other purposes where corrosion is of great importance. The effect of the occlusion of hydrogen in steel has been shown by a number of investigators to be very important and, under certain conditions, the volume of this gas occluded will reach nearly 50% of that of the metal itself.

The rate of hydrogen removal on depolarization depends upon solution pressures, overvoltage, catalytic properties, and upon relative potential differences of various parts of the surface—a complexity much greater than can be explained in the above simple manner.

It is common knowledge that two metals when alloyed often have greater resistance to corrosion than either component metal alone. The principle applies to any number of components. This may be due to the formation of some compound or compounds of the two metals, or under certain special conditions it might be due to the combination of one alloying metal with the impurities of the other metal, in such a manner as to cause the solution tension of the resulting compounds to be about alike. The effect of alloying a second metal with iron or steel may also effect the corrosion of the original iron or steel by increasing or decreasing the amount of occluded hydrogen.

Another important effect of the introduction of the second metal may be to form an oxide when corrosion commences which is of such an adherent nature as to form a firm coating, inhibiting further corrosion or preventing an excess of oxygen.

PURPOSE OF THE INVESTIGATION

Since it is impossible to be certain, *a priori*, what effect the addition of any metal will have on the properties of another, and as it has been shown that certain metals in small quantities improve the resistance of iron to atmospheric corrosion, these investigations were undertaken for the purpose of determining the effect of the addition of small quantities of cobalt on the atmospheric corrosion of iron and mild steel. We have particularly in mind the addition of small quantities of cobalt to the very pure iron prepared by the open-hearth method for sheet roofing material.

The comparative effects of small amounts of cobalt, nickel and copper were studied.

Our interest was stimulated by the positive nature of certain very early preliminary experiments, described in the next paragraph.

PRELIMINARY EXPERIMENTS

Very early in the autumn of 1912, in the course of these investigations on cobalt and its alloys, a preliminary set of alloys was prepared by adding small percentages of both cobalt and nickel to very pure iron. These alloys were exposed for several months on the roof of Nicol Hall, Queens University, Kingston, Ontario. After this exposure they were removed, and the amount of corrosion determined. In every case it was found that the addition of small percentages of cobalt and nickel had decreased the corrosion of the pure iron.[1]

Following this, a second set of alloys was made with like materials, in the same way, and exposed under similar conditions to the previous set, for a period from June 16, 1913, to October 16, 1913. At the end of this 122 days' exposure, the alloys were taken in, and the rate of corrosion in grams per square centimetre of exposed surface per year was computed.

Unfortunately, two of the alloys of this set met with an accident during the exposure; they were dropped from the supports, and came in contact with the metal roof, so that the series is not sufficiently complete to warrant giving all the details. However, the results were in general accord with those of the previous set, which led us to believe that the addition of cobalt in proper proportions to pure iron might prove of benefit to its non-corrosive properties.

The general method of procedure with these preliminary experiments was the same as that described in detail for the complete sets of experiments to be described below.

[1] The "pure" iron, was American ingot iron; see p. 5.

The two sets of experiments described above must be considered preliminary for a number of reasons, and primarily, because no heat treatment was given to the alloys.

Conclusions

1. From these preliminary experiments, it appears that additions of small percentages of both Co and Ni to American ingot iron add to its non-corrosive properties.

2. Cobalt seemed to be more effective than nickel when used in like amount.

3. These results were such as to stimulate further interest, but were not sufficiently complete or satisfactory to warrant drawing definite conclusions, particularly as to the relative effect of nickel and cobalt.

EXPERIMENTS ON NON-CORROSIVE ALLOYS PREPARED BY ADDITIONS OF COBALT, NICKEL, AND COPPER, TO AMERICAN INGOT IRON

Following are the data of three extended series of observations, from which our conclusions are drawn. The materials used for the preparation of the alloys of all three series, and in the general procedure also, were the same for all, and are given before the numerical corrosion values for the series.

Materials for Preparation of Alloys

The base of all the alloys was American ingot iron, furnished by Dr. Beck, American Rolling Mill Company, Middletown, Ohio. The analysis of this material, as furnished by Dr. Beck, is:—

	%
Fe	99.9
S	0.023
P	0.004
C	0.010
Mn	0.031
Cu	0.028
O	0.035
Si	trace
Ni	none
Ca	none

Our own analysis of the material is as follows:—

	%
Fe	99.9
S	0.027
P	0.0075
C	0.010
Mn	0.027
Cu	0.048
Si	trace
Ca	none
Ni	none

Inasmuch as there has been some discussion in recent literature as to the effect of copper in adding to the non-corrosive properties of American ingot iron and similar materials, we have made, in addition to the usual check analyses, additional analyses of the copper content. Checking our value of copper as 0·048, we have the following values from independent analyses:—

$$0·046$$
$$0·045$$
$$0·050$$

These were made by two independent analysts.

A later analysis of American ingot iron rolled into sheets for roofing material and shipped to this laboratory by the American Rolling Mills, Middletown, Ohio, is as follows:—

Sample No 34175[1]—American Ingot Iron Rosion Sheet, 8 ft. × 4 ft.

S	0·026
P	0·009
C	0·010
Mn	0·022
Cu	0·016

The cobalt, nickel, and copper, were of a correspondingly high degree of purity, analyzing, respectively, 99·7% cobalt, 99·3% nickel, and 99·8% copper. The cobalt was prepared in this laboratory by reduction of purified oxide,[2] and the copper and nickel were procured from reliable sources, and analyzed in this laboratory, as above stated.

METHOD OF PREPARATION OF ALLOYS

Crucible and Furnace

The alloys were all made in lined graphite crucibles, obtained either from the Dixon Crucible Company, or the Jonathan Bartley Crucible Company, and were either No. 3 or No. 5 size. These crucibles were lined with first-grade powdered magnesite, the magnesite being mixed with water, to bind it until set.

A Hoskins electric furnace of the carbon plate resistor type, was used for melting these alloys.

Melting and Casting

The components of the alloy to be prepared, were weighed out and put into the crucible together. These were often put into a cold furnace to be melted without pre-heating, while some were pre-heated in the Monarch Oil furnace, previous to putting them into the hot electric furnace.[3] The temperature of the inside of the furnace was measured from time to time, observations being made with a Wanner optical pyrometer.

After the melt had received what was considered a proper furnace treatment, powdered aluminium was added as a degasifier, and it was then poured into an iron mould of variable depth, which formed a cylindrical ingot about 1·25″ diameter. The alloys of Series I and II were cast in a square mould of about the same volume. The casting usually weighed in the neighbourhood of 2 pounds.

---

[1] Analysis furnished by Dr. Beck, American Rolling Mill Co., Middletown, Ohio.
[2] "Preparation of Metallic Cobalt by Reduction of the Oxide," by Herbert T. Kalmus, Report No. 259, Mines Branch, Dept. of Mines, Ottawa, 1913. Journal of Industrial and Engineering Chemistry, 1914, Vol. VI, p. 107.
[3] For a description of this furnace, see "Preparation of Metallic Cobalt by Reduction of the Oxide," Herbert T. Kalmus, Report No. 259, Mines Branch, Dept. of Mines, Ottawa, 1913.

The alloy was not considered satisfactory unless the crucible lining remained intact throughout, and after the furnace treatment showed no evidence of having allowed the melt to come in contact with the crucible itself.

Preparation of Discs for Corrosion Tests

The castings obtained were usually from 4" to 6" long, and about 1·25" diameter. To obtain samples for corrosion tests, these bars were turned into discs, and when satisfactory, the discs were annealed in a Fletcher Russell Muffle furnace.[1] For the small samples this was done by heating them at 780°C for two hours and allowing them to cool with the furnace during a period of six hours.

After turning off a skin to get rid of the outside layer, the disc was finished to a smooth surface in a lathe, the finishing chips being kept for analysis for carbon. After this operation the discs were carefully polished on the buffing lathe and measured for exposure.

Method of Measurement to Determine Corrosion

Before exposure, several careful measurements were made of the diameter and thickness of the discs, the average of the measurements being taken to compute the surface. In the first preliminary set of experiments the discs were carefully weighed, washed in alcohol, to remove any form of grease, and suspended by silk threads from pegs set in a large board. The board with samples was placed on the roof. In the second preliminary set of experiments, and in Series I to III, instead of suspending the discs by silk cords, holes were drilled in one edge of them, and vulcanized fibre pegs were tightly inserted. These were mounted on frames, as shown on Plates I and II. This method of support was resorted to as the silk cords proved unreliable; and on several occasions, after particularly severe weather conditions, samples were ruined by falling to the metal roof.

After corroding a sufficient length of time, the samples were brought in, and the rust removed with a 20% solution of ammonium citrate. They were then carefully washed in alcohol, dried, and weighed, and the loss in weight determined as shown in the tables to follow.

Absorption of Carbon Monoxide Gas

Although no carbon or carbon-containing materials were added to the charge, it was always found, upon analyzing the alloy, that the carbon content had increased; in one case from 0·010% to 0·48%. The conclusion seems to be that carbon was absorbed from the CO atmosphere within the furnace. The action did not seem to be uniform under the conditions that prevailed, and we did not determine the conditions which governed the rate of absorption of carbon by the metal. It was definitely observed that it was far more difficult to obtain low carbon castings of the cobalt-iron alloys than it was to obtain low carbon castings of either the nickel-iron or copper-iron alloys.

---

[1] Series I and II were not annealed; Series III, IV, and V were annealed.

PLATE I.

Corrosion tests: method of exposing alloy discs.

PLATE II.

Corrosion tests: method of exposing alloy discs.

CORROSION TESTS

SERIES I

The following alloys comprising Series I were exposed on the roof of Nicol Hall, as described above, beginning at 10 o'clock on the morning of March 18, 1914. They were removed on August 31, 1914, after an exposure of 3,984 hours.

Data of Corrosion Experiments: Series I

December 2, 1913.

ALLOY H 202
Proximate Analysis...Iron............99·75%
 Cobalt............0·25
Furnace..........Hoskins Resistor furnace, type F.C. No. 105.
Crucible.........Graphite No. 3, magnesite lined.
Soaking..........Duration............30 mins.
 Temperature............1500°C.
Maximum Temperature reached by melt....1560°C.
Degasifier..........Powdered aluminium...0·2 grams.
Heat Treatment.....None.

The crucible, furnace, heat treatment, soaking, and degasifier for the following set of alloys were identical with that given under alloy H 202.

February 10, 1914.

Alloy H 209
 Iron............99·5%
 Cobalt............0·5

November 25, 1913.

Alloy H 198
 Iron............99·0%
 Cobalt............1·0

November 25, 1913.

Alloy H 199
 Iron............98·0%
 Cobalt............2·0

November 26, 1913.

Alloy H 200
 Iron............97·0%
 Cobalt............3·0

November 29, 1913.

Alloy H 204
 Iron............99·75%
 Nickel............0·25

February 4, 1914.

Alloy H 207
 Iron............99·5%
 Nickel............0·5

November 22, 1913.

Alloy H 195
 Iron............99·0%
 Nickel............1·0

November 23, 1913.

Alloy H 196
 Iron............98·0%
 Nickel............2·0

SERIES I—*Continued.*

November 24, 1913.

Alloy H 197
- Iron................. 97·0%
- Nickel............... 3·0

November 28, 1913.

Alloy H 205
- Iron................. 99·75%
- Copper.............. 0·25

February 4, 1914.

Alloy H 206
- Iron................. 99·5%
- Copper.............. 0·5

December 2, 1913.

Alloy H 208
- Iron................. 99·0%
- Copper.............. 1·0

Alloy H 216
- Iron................. 100·0%

There are also included in this set, several alloys made up under mill conditions by the American Rolling Mill Company, having the following compositions:—

Alloy 34204
- Co................... 0·35%
- C.................... 0·010
- Mn................... 0·020
- P.................... 0·009
- S.................... 0·022
- Cu................... 0·020

Alloy 34196
- Co................... 0·60%
- C.................... 0·010
- Mn................... 0·020
- P.................... 0·008
- S.................... 0·040
- Cu................... 0·024

Alloy 34185
- Co................... 1·18%
- C.................... 0·015
- Mn................... 0·017
- P.................... 0·006
- S.................... 0·034
- Cu................... 0·028

Alloy 44009
- Ni................... 0·75%
- C.................... 0·010
- Mn................... 0·015
- P.................... 0·008
- S.................... 0·025
- Cu................... 0·24

Alloy 34175
- C.................... 0·010%
- Mn................... 0·022
- P.................... 0·009
- S.................... 0·026
- Cu................... 0·016

SERIES I—*Continued.*

These alloys, including the one sample of iron No. 34175, received no heat treatment after coming into our hands in the form of partially rolled sheets, approximately ½" thick, by 10" wide. The discs were cut directly from these sheets, and after the proper measurements, were exposed in the manner already described for Series I.

The following table shows the corrosion of these various alloys:

Table of Particulars of Corrosion Tests of Alloys Utilized in Series I

Sample number of alloy	Approximate analysis	Diameter in cms.	Thickness in cms.	Area in square cms.	Weight in grams before exposure	Weight in grams after removal of rust	Loss of weight in grams due to corrosion	Corrosion or loss in weight in grams per sq. cm. of original surface per hr. × 10⁶	Remarks

Exposed March 18, 1914, ten o'clock. Removed August 31, 1914. Exposure 3,984 hours.

H 204 (a)	Fe 99·75, 0·25	3·0053	0·5245	19·60	31·7434	31·4464	0·3874	752	Rust quite tenacious.
H 204 (b)		2·9353	0·4839	17·99	25·4480	24·9393	0·4944	696	
H 207 (a)	Fe 99·5, Ni 0·50	2·9972	0·5167	18·98	28·3184	27·9340	0·3844	510	Rust quite tenacious.
H 207 (b)		2·6507	0·5299	15·45	22·7629	22·3216	0·4413	717	
H 207 (c)		2·4244	0·4435	12·96	15·9782	15·5042	0·4740	905	
H 195 (a)	Fe 99·0, Ni 1·	2·9420	0·6670	19·43	44·7888	44·2410	0·5478	694	Rust tenacious. Sample (b) had fallen off.
H 195 (b)		2·9126	0·6346	19·43	33·0013	32·2974	0·7069	328	
H 196 (a)	Fe 98·0, Ni 2·0	2·9515	0·6293	19·52	33·6308	33·1004	0·5304	681	Rust removed somewhat easier than H 195.
H 196 (b)		2·9538	0·6743	19·96	36·0536	35·5070	0·5466	687	
H 197 (a)	Fe 97·0, Ni 3·0	2·9940	0·5960	19·68	33·4942	32·9300	0·4642	588	Rust easily removed.
H 197 (b)		2·9786	0·4416	18·036	23·9081	23·4407	0·4674	652	
H 202 (a)	Fe 99·75, Co 0·25	2·9068	0·3958	16·89	20·4681	20·0148	0·4533	673	Rust removes easier than H 200. Sample (b) had fallen off.
H 202 (b)		3·0148	0·3898	17·95	21·5152	20·9614	0·5543	775	
H 209 (a)	Fe 99·5, Co 0·5	2·7904	0·5974	17·47	28·4904	27·9386	0·5518	793	Both samples off. Rust removed quite easily. Sample (b) had not rusted much on one side.
H 209 (b)		2·9556	0·5765	19·08	30·5810	30·2152	0·3658	378	
H 198 (a)	Fe 99·0, Co 1·0	2·9875	0·6615	20·172	36·2452	35·6630	0·5822	722	Rust dark in colour and more tenacious than H 197.
H 198 (b)		2·9629	0·6225	19·685	34·3751	34·0090	0·3661	520	

SERIES I—Continued.

Dimensions of Disc before Exposure   *Weight of Disc*

Sample number of alloy	Approximate analysis	Diameter in cms.	Thickness in cms.	Area in square cms.	Weight in grams before exposure	Weight in grams after removal of rust	Loss of weight in grams due to corrosion	Corrosion or loss in weight in grams per sq. cm. of original surface per hr. × 10⁸	Remarks
H 199 (a)	Fe 98·0 % Cu 2·0	2·7845	0·6128	17·422	28·5496	28·0173	0·5323	774	Rust dark in colour and tenacious.
H 199 (b)		2·9315	0·5527	18·590	29·2209	28·6635	0·5574	768	Rust dark in colour and quite tenacious.
H 200 (a)	Fe 97·0 % Cu 3·0	3·0368	0·5593	19·82	41·5005	41·1169	0·3836	482	Rust dark in colour and quite tenacious.
H 200 (b)		4·0128	0·5533	19·85	41·0141	30·5239	0·4882	627	
H 205 (a)	Fe 99·75 % Cu 0·25	2·8759	0·5133	17·90	27·5937	27·0692	0·5245	748	Rust about the same as H 204.
H 205 (b)		2·8853	0·5895	18·42	30·1322	29·5943	0·5379	744	
H 206 (a)	Fe 99·5 % Cu 0·50	2·9452	0·5497	18·73	29·2453	28·6726	0·5677	761	Rust about the same as H 204.
H 206 (b)		2·9439	0·5879	18·81	29·6187	29·0536	0·5501	748	
H 208 (a)	Fe 99·0 % Cu 1·0	2·3845	0·7078	14·23	24·3692	23·9062	0·4630	816	Rust removed rather easily.
H 208 (b)		2·3772	0·6475	13·71	22·4719	21·9154	0·5365	835	
H 216 (a)	Fe 100·0 %	4·6021	0·3611	26·73	44·8356	43·8023	1·0733	968	Rust quite easily removed. Sample (a) had fallen off.
H 216 (b)		3·7567	0·6153	29·43	53·3836	52·2870	1·0966	937	
34175 (a)	S 0·027 % Mn 0·077 P 0·078 Cu 0·020 C 0·131	3·769	1·118	35·47	97·6381	96·7020	0·9351	664	Rust removed very easily.
34175 (b)		3·683	1·118	34·18	92·2952	92·3370	0·9582	701	Rust removed very easily.
34204 (a)	Co 0·22 % S 0·022 Mn 0·046 P 0·0058 Cu 0·020 C 0·125	3·718	1·047	43·05	89·1774	87·5760	1·6014	1,180	Rust more tenacious than 34175, but less tenacious than 34185.
34204 (b)		3·698	1·049	33·32	88·3738	87·9770	0·3968	292	
34196 (a)	Co 0·57 % S 0·025 Mn 0·031 P 0·0097 Cu 0·023 C 0·136	3·723	1·101	44·32	94·0772	93·2020	0·8752	649	Rust tenacious; not very different from 34204.
34196 (b)		3·735	1·145	45·22	97·3638	96·5120	0·9518	678	
34185 (a)	Co 1·09 % S 0·027 Mn 0·032 P 0·008 Cu 0·025 C 0·148	3·682	1·105	34·20	92·2378	91·1260	1·1118	823	Rust very tenacious.

Exposed March 18, 1914, ten o'clock.   Removed August 31, 1914.   Exposure 3,984 hours.

SERIES I—Continued.

Sample number of alloy	Approximate analysis	Diameter in cms.	Thickness in cms.	Area in square cms.	Weight before exposure in grams	Weight after removal of rust in grams	Loss of weight due to corrosion in grams	Corrosion or loss in weight in grams per sq. cm. of original surface per hr. × 10⁶	Remarks
34185 (b)		3.712	1.091	34.22	32.5529	31.7200	0.8329	613	Rust very tenacious.
44009 (a)	Ni 0.70 S 0.020 Mn 0.025 P 0.0065 Cu 0.12 C 0.140	3.728	1.014	34.75	86.8718	85.8190	1.0528	783	Rust very tenacious, but unlike 34185.
44009 (b)		3.739	1.063	34.50	91.4208	90.2970	1.1238	819	

Microphotographs. No microphotographs were taken of the alloys of the foregoing Series I.

Conclusions

Corrosion Experiments, Series I

(1.) Results with the set of alloys, numbers 196 to 216, show that in every case the alloy formed by the addition of cobalt, nickel, or copper, is less corroded in the atmosphere than is American ingot iron. These are unannealed samples.

(2.) The variations in the measurements are not so great but that conclusion 1 is apparent, but they are sufficiently great that conclusions based upon a comparison of the alloys among themselves could not be drawn from this series alone.

(3.) The samples, numbers 34175 to 44009, prepared by the American Rolling Mill Company, show such wide variations among the observations, that no conclusions could be drawn from this series alone.

(4.) It is noticeable throughout this series, that the rust on the cobalt samples is more tenacious than that of the other samples, and particularly that it is of a much darker colour, and, is removed, by mechanical means, with very much greater difficulty than that formed on the American ingot iron.

Corrosion Tests

SERIES II

The alloys H 195 to H 216 used for Series II were identical with those used for Series I. That is to say, after completing the measurements for Series I, the discs were again polished, re-weighed, and measured for a second exposure.

In this second series, the American Rolling Mill alloys of the first series, were again exposed, but in this case the samples were not the identical ones used in the previous test, new samples being cut from the sheets, and prepared for exposure in the manner described.

The frame of samples was placed on the Nicol building of Queens University, Kingston, Ontario, on the morning of October 10, 1914, and removed on August 30, 1915.

The results of this second exposure may be noted in the following table:—

Table of Particulars of Corrosion Tests of Alloys Utilized in Series II

Sample number of alloy	Approximate analysis	Diameter in cms.	Thickness in cms.	Area in square cms.	Weight in grams before exposure	Weight in grams after removal of rust	Loss of weight in grams due to corrosion	Corrosion or loss in weight in grams per sq. cm. of original surface per hr. × 10⁸	Remarks
		Exposed October 10th, 1914.			Removed August 30, 1915.				
H 204 (a)	Fe 99.75 Ni 0.25	2.984	0.540	17.59	28.9656	28.2924	0.6732	491	Rust difficult to remove; dark in spots and mottled in appearance.
H 204 (b)		2.918	0.453	16.70	22.9425	22.3325	0.6100	470	
H 207 (a)	Fe 99.5 Ni 0.5	2.080	0.487	18.49	26.4491	25.5458	0.6033	420	Rust fairly tenacious.
H 207 (b)		2.621	0.481	13.66	30.0225	19.5246	0.4979	450	
H 207 (c)		2.480	0.401	12.76	14.9802	14.9915	0.3987	402	
H 195 (a)	Fe 99.0 Ni 1.0	2.898	0.600	18.25	With wax in holes 30.0960 Without wax 30.0200	29.2932	0.7268	512	Rust tenacious; somewhat dark after removal of rust.
H 195 (b)		2.893	0.371	18.53	29.0754	28.4344	0.6410	445	
H 196 (a)	Fe 98.0 Ni 2.0	2.938	0.558	19.08	30.9354	30.2866	0.6488	447	Rust easily removed; light in colour.
H 196 (b)		2.940	0.611	19.43	33.7160	33.0554	0.6606	437	
H 197 (a)	Fe 97.0 Ni 3.0	2.983	0.561	19.48	30.2280	29.6297	0.5983	401	Rust easily removed; light in colour.
H 197 (b)		2.961	0.412	17.58	21.8434	21.2940	0.5494	402	

SERIES II—*Continued*

Dimensions of Disc before Exposure — *Weight of Disc*

Sample number of alloy	Approximate analysis	Diameter in cms.	Thickness in cms.	Area in square cms.	Weight in grams before exposure	Weight in grams after removal of rust	Loss of weight in grams due to corrosion	Corrosion or loss in weight in grams per sq. cm. of original surface per hr. × 10⁷	Remarks
		Exposed October 10, 1914.			Removed August 30, 1915.				
H 202 (a)	Fe 97.75% Co 0.25	2.885	0.371	16.14	18.7366	18.2327	0.4939	486	Rust dark, tenacious, and uniform over surface.
H 209 (a)	Fe 99.5% Co 0.5	2.759	0.550	16.72	25.3148	24.8650	0.6495	445	Rust medium dark in colour.
H 209 (b)		2.938	0.534	18.51	27.7878	27.1625	0.6253	500	
H 198 (a)	Fe 99.0% Co 1.0	2.965	0.616	19.59	33.0186	32.2596	0.7590	498	Rust dark in colour; fairly difficult to remove.
H 198 (b)		2.940	0.590	19.04	30.5180	29.8162	0.7018	475	
H 199 (a)	Fe 98.0% Co 2.0	2.730	0.584	16.74	26.4260	25.7902	0.6364	489	Medium dark in colour.
H 199 (b)		2.911	0.522	18.09	26.8364	26.1593	0.6771	482	
H 200 (a)	Fe 97.0% Co 3.0	3.022	0.535	19.45	28.4286	27.8893	0.5393	356	Rust easily removed; metal underneath dark in colour.
H 200 (b)		3.002	0.528	19.08	28.8106	28.2609	0.5497	371	
H 205 (a)	Fe 99.75% Cu 0.25	2.854	0.516	17.38	25.1454	24.5636	0.5718	414	Rust hard in spots, metal underneath dark in colour.
H 205 (b)		2.865	0.557	17.93	27.4168	26.8236	0.5932	415	
H 206 (a)	Fe 99.5% Cu 0.5	2.918	0.515	18.07	26.6298	26.0090	0.6208	433	Rust light in colour; metal underneath pitted.
H 206 (b)		2.918	0.520	18.12	26.8750	26.2528	0.6222	432	
H 208 (a)	Fe 99.0% Cu 1.0	2.333	0.674	13.65	21.3946	21.0834	0.3112	287	Rust very uniformly distributed and difficult to remove.
H 208 (b)		2.350	0.640	13.20	20.4918	20.0884	0.4034	392	
H 216 (a)	Fe 100.0%	3.578	0.522	25.97	40.7980	39.5724	1.2256	609	Rust easily removed.
H 216 (b)		3.732	0.584	28.60	49.7338	48.1793	1.5545	698	
34175 (a)	S 0.027% Mn 0.427 P 0.0078 Cu 0.020 C 0.131	2.897	0.769	20.12	36.7010	35.7888	0.9122	584	Rust light in colour and fairly easily removed.
34175 (b)		2.561	0.768	16.42	30.5996	29.8434	0.7862	592	Rust light in colour and fairly easily removed.
34175 (c)		2.293	0.770	15.70	24.5528	23.9235	0.6293	591	

SERIES II—Continued.

Sample number of alloy	Approximate analysis	Diameter in cms.	Thickness in cms.	Area in square cms.	Weight in grams before exposure	Weight in grams after removal of rust	Loss of weight in grams due to corrosion	Corrosion or loss in weight in grams per sq. cm. of original surface per hr. × 10⁸	Remarks
		Dimensions of Disc before Exposure			*Weight of Disc*				
		Exposed October 10, 1914. Removed August 30, 1915.							
34173 (d)		2·080	0·766	11·47	19·4738	18·8458	0·5480	605	
34204 (a)	Co 0·22 / S 0·022 / Mn 0·036 / P 0·0058 / Cu 0·020 / C 0·125	4·952	0·767	44·00	73·1770	71·6781	1·4989	567	Rust light in colour and fairly easily removed.
34204 (b)		3·700	0·774	30·31	64·5962	63·1962	1·4000	594	
34204 (c)		3·430	0·768	26·73	55·1002	53·9060	1·1942	604	
34204 (d)		3·180	0·766	23·02	47·2278	46·1447	1·0831	562	
34196 (a)	Co 0·57 / S 0·025 / Mn 0·041 / P 0·0097 / Cu 0·021 / C 0·156	2·930	0·769	20·50	40·2700	39·3240	0·9461	593	Rust light in colour and fairly easily removed.
34196 (b)		2·672	0·766	17·56	33·1640	32·3249	0·8391	613	
34196 (c)		2·438	0·766	15·15	27·7012	26·9839	0·7173	609	
34196 (d)		2·171	0·763	12·53	21·7768	21·2004	0·5764	592	
34185 (a)	Co 1·09 / S 0·027 / Mn 0·032 / P 0·008 / Cu 0·025 / C 0·145	4·810	0·774	32·03	68·4924	67·1646	1·3278	534	Rust light in colour and fairly easily removed.
34185 (b)		3·553	0·774	28·48	59·6462	58·4286	1·2176	551	Rust light in colour and fairly easily removed.
34185 (c)		3·318	0·770	25·22	51·6070	50·5487	1·0583	539	
34185 (d)		3·068	0·769	22·14	44·3810	43·4005	0·9805	570	
44009 (a)	Ni 0·700 / S 0·020 / Mn 0·025 / P 0·0065 / Cu 0·270 / C 0·140	3·515	1·025	27·44	69·1338	68·0767	1·0571	779	Rust light in colour and fairly easily removed.
44009 (b)		3·065	1·027	24·95	58·9434	58·0252	0·9182	788	
44009 (c)		2·812	1·032	17·64	49·8336	48·9872	0·8464	860	
44009 (d)		2·560	1·026	10·47	41·0208	40·3029	0·7179	883	

Microphotographs.—No microphotographs were taken of Series II.

Conclusions.

Corrosion Experiments, Series II.

(1.) Results with the set of alloys, numbers 196 to 216, show that in every case the alloy formed by the addition of cobalt, nickel, or copper, is less corroded in the atmosphere than is American ingot iron. These are unannealed samples.

(2.) The conclusion 1, for Series II, is in accord with the corresponding conclusion for Series I. Comparing the absolute amounts of corrosion, that is, loss in weight in grams per square centimetre of original surface per hour, we find that it is uniformly greater for Series I than for Series II—approximately in the ratio of three to two. This may be in some measure accounted for by the fact that the exposure for Series I was largely in the summer time, whereas the exposure for Series II was in both summer and winter. It is more to be accounted for, however, by the fact that the exposure for Series II was of very nearly twice the duration that it was for Series I. After corrosion has continued for a certain period, especially with the more non-corrosive alloys, a hard, tenacious dark-coloured rust is formed, which tends to protect the alloy against further corrosion.

(3.) The variations in the measurements are not so great but that conclusions 1 and 2 are apparent, but the variations are sufficiently great that comparisons between the various alloys cannot be drawn without further confirming measurements to establish the relations.

(4.) The samples, numbers 34175 to 44009, prepared by the American Rolling Mill Company—as far as this series alone is concerned—do not tend to bear out the conclusions from the series prepared at the laboratory. From these samples alone it would seem that additions of small amounts of cobalt, up to one per cent, have very little effect on corrosion, and that the addition of nickel to about 0·7 of a per cent was harmful.

(5.) It is noticeable throughout this series, in the samples prepared at the laboratory, that additions of cobalt, nickel, and copper, all tend to make the rust more tenacious, darker in colour, more uniform, and removed by mechanical means with very much greater difficulty than in the case of the pure American ingot iron. This fact is particularly noticeable with the cobalt samples. The alloys of this series, prepared by the American Rolling Mill Company, differ from those prepared by us; that is to say, the annealed samples differ from the unannealed samples, in that for the annealed samples the rust is light in colour, and much more readily removed than in the case of the unannealed ones.

(6.) From this series alone we would tend to conclude that the effect of annealing these alloys was to promote corrosion.

SERIES III.

The samples of Series I and II received no heat treatment after casting. For further investigation Series III was prepared, the following being a list of the alloys of this series, with the method of preparation.

PREPARATION OF ALLOYS

Furnace Treatment

In this series of alloys, some of the charges were preheated in the Monarch Oil Furnace, and some were introduced directly into the hot Hoskins furnace. The data of preparation of two alloys by each treatment are given in detail. All samples of this series were prepared by one or the other of these methods as indicated. Any variation from these will be noted under the individual charge.

DATA OF CORROSION EXPERIMENTS

October 23, 1914.

Alloy B 202
- Composition......... Co — 0·25%; C — 0·18%; Fe — 99·5%.
- Charge............... American ingot iron........ 1132 grams
  - Cobalt....................... 2·85 "
- Furnace.............. Hoskins Resistor furnace, type F.C. No. 105.
- Crucible............. No. 3 graphite, magnesite lined.
- Soaking.............. 30 minutes at 1570° C.
- Highest Temperature reached by melt 1570° C.
- Temperature of Pouring 1570° C.
- Degasifier........... Powdered aluminium, 0·2 grams.
- Weight of Casting... 1 lb. 4 oz.
- Heat Treatment...... Annealed by heating in gas muffle furnace at 870° for 2 hrs, and allowing to cool slowly with furnace.
- Microphotograph..... (See Plate III, p. 30).

November 11, 1914.

Alloy B-199
- Composition......... Co — 2·0%; C — 0·16%; Fe — 97·8%.
- Charge............... S 262........................ 603 grams
  - Cobalt....................... 6·15 "
- Soaking 30 minutes at 1640° C.
- Highest Temperature reached by melt 1700° C.
- Temperature of Pouring 1640° C.
- Crucible, Furnace Treatment, Degasifier, and Heat Treatment same as B 202, except as noted.
- Weight of Casting... 1 lb.
- Microphotograph (See Plate IV, p. 30).

November 2, 1914.

Alloy S 250
- Composition......... Co — 0·25%; C — 0·083%; Fe — 99·6%.
- Charge............... American ingot iron........ 778 grams
  - Cobalt....................... 1·95 "
- Furnace.............. Hoskins Resistor furnace, type F.C. No. 105 Monarch Oil furnace.
- Crucible............. No. 3 graphite, magnesite lined.
- Preheating.......... 20 minutes in oil furnace.

SERIES III—Continued.

Soaking .... 30 minutes at 1560° C.
Highest Temperature reached by melt, 1640° C.
Temperature of Pouring 1560° C.
Degasifier .... Powdered aluminium, 0.2 grams.
Weight of Casting ..... 1 lb. 3 oz.
Heat Treatment .... Annealed by heating in gas muffle furnace at 870° C. for 2 hours and allowing to cool slowly with furnace.
Microphotograph .... (See Plate V, p. 30).

November 10, 1914.

Alloy S 262
Composition .... Co = 1.0%, ; C = 0.62% ; Fe = 98.3%.
Charge .... S 260   456 grams.
                         S 255   226   "
                         B 200   130   "
                         Pure Cobalt   0.49   "
Preheating .... 30 minutes in oil furnace.
Soaking .... 30 minutes at 1640° C.
Highest Temperature reached by melt, 1700° C.
Temperature of Pouring 1640° C.
Crucible, Furnace Treatment, Degasifier, and Heat Treatment same as S 250, except as noted.
Weight of Casting ..... 1 lb. 5 ozs.
Microphotograph .... (See Plate VI, p. 30).

November 9, 1914.

Alloy S 255
Composition .... Co = 0.35%, ; C = 0.21% ; Fe = 99.4%.
Charge .... S 252 .... 375.5 grams.
                         American ingot iron. 504   "
                         Cobalt ....1.53   "
Crucible, Furnace Treatment, Degasifier, and Heat Treatment same as S 262.
Microphotograph .... (See Plate VII, p. 30).
NOTE. This alloy was cast four times before obtaining a satisfactory sample.

December 9, 1914.

Alloy S 265
Composition .... Co = 0.35%, ; C = 0.30% ; Fe = 99.3%.
Charge .... Alloy 32404   467 grams.
                         Cobalt   0.16 "
Crucible, Furnace Treatment, Degasifier, Weight of Casting, and Heat Treatment same as B 199, with exception that duration of soaking was 1 hour.

October 26, 1914.

Alloy B 209
Composition .... Co = 0.50%, ; C = 0.27% ; Fe = 99.2%.
Charge .... B 202.   637 grams.
                         Cobalt .... 1.54 "
Crucible, Furnace Treatment, Degasifier, and Heat Treatment same as B 202.
Weight of Casting ..... 1 lb.
Microphotograph .... (See Plate VIII, p. 30).

SERIES III—Continued.

November 2, 1914.

Alloy S 282
Composition......... Co — 0·50%; C — 0·31%; Fe — 99·1%.
Charge............. S 250............................ 650 grams.
                    Cobalt........................... 1·55 "
Crucible, Furnace Treatment, Degasifier, and Heat Treatment same as S 250.
Weight of Casting... 1 lb. 2 ozs.
Microphotograph.... (See Plate IX, p. 30).

November 9, 1914.

Alloy S 260
Composition......... Co — 0·75%; C — 0·17%; Fe 99·1%.
Charge............. American ingot iron............. 576 grams.
                    B 200............................ 192 "
Crucible, Furnace Treatment, Degasifier, and Heat Treatment same as S 262.
Weight of Casting... 1 lb. 5 ozs.
Microphotograph.... (See Plate X, p. 30).

December 9, 1914.

Alloy S 266
Composition......... Co — 0·75%; C — 0·21%; Fe — 99·0%.
Charge............. Alloy 34204...................... 467 grams.
                    Cobalt........................... 2·35 "
Crucible, Furnace Treatment, Degasifier, and Heat Treatment same as B₂ 199, with exception that duration of soaking was 2 hours.
Weight of Casting... 1 lb.
NOTE.—This alloy was cast twice before obtaining a satisfactory sample.

October 27, 1914.

Alloy B 198
Composition......... Co — 1·0%; C — 0·38%; Fe — 98·6%.
Charge............. B 202............................ 326 grams.
                    B 209............................ 460 "
                    Cobalt........................... 4·80 "
Crucible, Furnace Treatment, Degasifier, and Heat Treatment same as B 202.
Weight of Casting... 1 lb. 8 ozs.
Microphotograph.... (See Plate XI, p. 30).

November 13, 1914.

Alloy B₂ 200
Composition......... Co — 3·0%; C — 0·17%; Fe — 96·8%.
Charge............. B₂ 199........................... 340 grams.
                    American ingot iron.............. 262 "
                    Cobalt........................... 11·63 "
Crucible, Furnace Treatment, Degasifier, and Heat Treatment same as B₂ 199.
Weight of Casting... 1 lb. 2 ozs.
NOTE.—This alloy was cast twice before obtaining a satisfactory sample.
Microphotograph.... (See Plate XII, p. 30).

SERIES III—Continued.

November 29, 1914.

Alloy B 200
Composition........ Co = 3·0%; C = 0·36%; Fe 96·6%.
Charge............. B 199........................ 482 grams.
                    Cobalt....................... 1·49 "
Crucible, Furnace treatment, Degasifier, and Heat Treatment same as B 202, except highest temperature was 1680° C.
Weight of Casting... 1 lb.

November 25, 1914.

Alloy C 202
Composition........ Co = 0·25%; C = 0·49%; Fe 99·2%.
Charge............. B 30......................... 255 grams.
                    American ingot iron........... 255 "
                    Cobalt....................... 1·28 "
Crucible, Furnace Treatment, Degasifier, and Heat Treatment same as B 199, with the exception that duration of soaking was 50 minutes.
Weight of Casting... 1 lb.
Microphotograph.... (See Plate XIII, p. 30).

November 26, 1914.

Alloy C 255
Composition........ Co = 0·35%; C = 0·49%; Fe = 99·2%.
Charge............. C 202........................ 305 grams.
                    B 30.......................... 85 "
                    Alloy 32404................... 85 "
                    Cobalt....................... 0·69 "
Crucible, Furnace Treatment, Degasifier, and Heat Treatment same as B 199, with the exception that duration of soaking was 1 hour.
Weight of Casting... 15 ozs.
Microphotograph.... (See Plate XIV, p. 30).

October 28, 1914.

Alloy B 199
Composition........ Co = 2·0%; C = 0·46%; Fe = 97·5%.
Charge............. B 198........................ 651 grams.
                    Cobalt....................... 0·65 "
Crucible, Furnace Treatment, Degasifier, and Heat Treatment same as B 202.
Weight of Casting... 1 lb. 5 ozs.
Microphotograph.... (See Plate XV, p. 30).

October 23, 1914.

Alloy B 204
Composition........ Ni = 0·25%; C = 0·10%; Fe = 99·6%.
Charge............. American ingot iron........... 1132 grams.
                    Pure Nickel................... 2·85 "
Crucible, Furnace Treatment, Degasifier, and Heat Treatment same as B 202.
Weight of Casting... 1 lb. 4 ozs.

SERIES III—Continued.

November 2, 1914.

Alloy S 25i

Compo. .... Ni — 0·25%; C — 0·057%; Fe — 99·6%
Charge.......... American ingot iron...... 743 grams.
                 Nickel............... 1·86 "
Crucible, Furnace Treatment, Degasifier, and Heat Treatment same as S 250.
Weight of Casting. 1 lb. 4 ozs.
Microphotograph. (See Plate XVI, p. 30).

November 5, 1914.

Alloy S 284

Composition ..... Ni — 0·35%; C — 0·045%; Fe — 99·6%
Charge.......... American ingot iron...... 913 grams
                 Nickel............... 3·19 "
Crucible, Furnace Treatment, Degasifier, and Heat Treatment same as S 262, except that highest temperature was 1640°C.
Weight of Casting. 1 lb.
Note.—This bar was cast seven times before obtaining a perfect casting.
Microphotograph. (See Plate XVII, p. 30).

November 6, 1914.

Alloy S 257

Composition ..... Ni — 0·35%; C — 0·060%; Fe — 99·6%
Charge.......... S 254................. 226 grams.
                 American ingot iron.. 495 "
                 Nickel............... 1·72 "
Crucible, Furnace Treatment, Degasifier, and Heat Treatment same as S 250.
Weight of Casting. 1 lb. 3 ozs.
Note.—This alloy was cast twice before obtaining a satisfactory sample.

November 6, 1914.

Alloy S 258

Composition..... Ni — 0·50%; C — 0·072%; Fe — 99·4%
Charge.......... American ingot iron... 935 grams.
                 Nickel............... 4·67 "
Crucible, Furnace Treatment, Degasifier, and Heat Treatment same as S 262.
Weight of Casting. 1 lb. 5 ozs.
Note.—This alloy was cast three times before obtaining a satisfactory sample.
Microphotograph. (See Plate XVIII, p. 30).

November 9, 1914.

Alloy S 259

Composition...... Ni — 0·75%; C — 0·067%; Fe — 99·2%
Charge.......... S 258................. 637 grams.
                 Nickel............... 1·60 "
Crucible, Furnace Treatment, Degasifier, and Heat Treatment same as S 262.
Weight of Casting. 1 lb. 5 ozs.
Microphotograph. (See Plate XIX, p. 30).

SERIES III—Continued.

November 19, 1914.

Alloy B495

Composition.... Ni – 1·0%; C – 0·089%; Fe – 98·9%.
Charge........ S 261.................. 177 grams.
                 Nickel................. 160   "
Crucible, Furnace Treatment, Degasifier, and Heat Treatment same as B₂ 199.
Weight of Casting... 1 lb.
Microphotograph.... (See Plate XX, p. 30).

November 10, 1914.

Alloy S 263

Composition.... Ni – 1·0%; C – 0·065%; Fe – 98·9%.
Charge........ S 261.................. 637 grams.
                 Nickel................. 6·50 "
Crucible, Furnace Treatment, Degasifier, and Heat Treatment same as B₂ 199 with the exception that duration of soaking was 45 minutes.
Weight of Casting... 1 lb. 3 ozs.
NOTE.—This alloy was cast twice before obtaining a satisfactory sample.
Microphotograph.... (See Plate XXI, p. 30).

December 11, 1914

Alloy S 269

Composition.... Ni – 2·0%; C – 0·085%; Fe – 97·9%.
Charge........ American ingot iron...... 566 grams.
                 Nickel................. 11·58 "
Crucible, Furnace Treatment, Degasifier, and Heat Treatment same as B₂ 199, except soaking was 1 hour at 1700° C.
Weight of Casting... 1 lb. 2 ozs.
Microphotograph.... (See Plate XXII, p. 30).

December 10, 1914.

Alloy S 267

Composition.... Ni – 0·25%; C – 0·23%; Fe – 99·5%.
Charge........ C 204................... 326 grams.
                 American ingot iron...... 435   "
                 Nickel................. 1·09 "
Crucible, Furnace Treatment, Degasifier, and Heat Treatment same as B₂ 199, except that soaking was 1 hour at 1700° C.
Weight of Casting... 1 lb. 8 ozs.

December 10, 1914.

Alloy S 268

Composition.... Ni – 0·50%; C – 0·21%; Fe – 99·2%.
Charge........ B 30................... 127 grams.
                 American ingot iron...... 467   "
                 Nickel................. 2·99 "
Crucible, Furnace Treatment, Degasifier, and Heat Treatment same as B₂ 199, except that soaking was 1 hour.
Weight of Casting... 1 lb. 6 ozs.

SERIES III—Continued.

October 24, 1914.

Alloy B 195

Composition......Ni − 1·0%; C − 0·24%; Fe − 98·8%.
Charge..........B 207......................800 grams.
                 Nickel.....................4·04 "
Crucible, Furnace Treatment, Degasifier, and Heat Treatment same as B 202.
Weight of Casting...1 lb. 4 ozs.
Microphotograph....(See Plate XXIII, p. 30).

October 27, 1914.

Alloy B 196

Composition......Ni − 2·0%; C − 0·23%; Fe − 97·8%.
Charge..........B 195......................680 grams.
                 Nickel.....................6·95 "
Crucible, Furnace Treatment, Degasifier, and Heat Treatment same as B 202, except highest temperature was 1640° C.
Weight of Casting...1 lb. 6 ozs.
Microphotograph....(See Plate XXIV, p. 30).

November 12, 1914.

Alloy B₂ 197

Composition......Ni − 3·0%; C − 0·13%; Fe − 96·8%.
Charge..........S 263......................184 grams.
                 B 197......................368 "
                 Nickel.....................1·90 "
Crucible, Furnace Treatment, Degasifier, and Heat Treatment same as B₂ 199.
Weight of Casting...8 ozs.
Microphotograph....(See Plate XXV, p. 30).

October 28, 1914.

Alloy B 197

Composition......Ni − 3·0%; C − 0·21%; Fe − 96·7%.
Charge..........B 196......................517 grams.
                 Nickel.....................5·33 "
Crucible, Furnace Treatment, Degasifier, and Heat Treatment same as B 202.
Weight of Casting...1 lb. 1 oz.
Microphotograph....(See Plate XXVI, p. 30).

November 25, 1914.

Alloy C 204

Composition......Ni − 0·25%; C − 0·43%; Fe − 99·3%.
                 B 30......................269·5 grams.
Charge..........American ingot iron........269·5 "
                 Nickel.....................1·35 "
Crucible, Furnace Treatment, Degasifier, and Heat Treatment same as S 250, except duration of soaking was 1 hour at 1560° C.
Weight of Casting...1 lb. 2 ozs.
Microphotograph....(See Plate XXVII, p. 30).

SERIES III—Continued.

November 16, 1914.

Alloy B₂ 205
Composition .... Cu — 0.25%; C — 0.045%; Fe — 99.7%.
Charge .... American ingot iron .... 1·062 grams.
                       Copper .... 2·66 "
Crucible, Furnace Treatment, Degasifier, and Heat Treatment same as B₂ 199.
Weight of Casting ... 2 lbs.
Note.—In this and the succeeding copper alloys the iron was first melted, and the copper dropped into the molten iron through a small hole in the furnace cover, the fall being directed by means of a tube of fused silica.
Microphotograph. (See Plate XXVIII, p. 30).

October 28, 1914.

Alloy B 205
Composition .... Cu — 0.25%; C — 0.19%; Fe — 99.6%.
Charge .... American ingot iron .... 793 grams.
                       Copper .... 0·25 "
Crucible, Furnace Treatment, Degasifier, and Heat Treatment same as B 202.
Weight of Casting ... 1 lb. 5 ozs.

November 19, 1914.

Alloy B₂ 206
Composition .... Cu — 0.50%; C — 0.17%; Fe — 99.3%.
Charge .... B₂ 205 .... 552 grams.
                       Copper .... 1·39 "
Crucible, Furnace Treatment, Degasifier, and Heat Treatment same as B 199.
Weight of Casting ... 1 lb. 2 ozs.
Microphotograph. See Plate XXIX, p. 30).

October 30, 1914.

Alloy B 206
Composition .... Cu — 0.50%; C — 0.19%; Fe — 99.3%.
Charge .... B 205 .... 488 grams.
                       Copper .... 1·23 "
Crucible, Furnace Treatment, Degasifier, and Heat Treatment same as B₂ 199, except that highest temperature was 1640° C.
Weight of Casting ... 1 lb.
Microphotograph. (See Plate XXX, p. 30).

November 24, 1914.

Alloy B 208
Composition .... Cu — 0.75%; C — 0.18%; Fe — 99.0%.
Charge .... B₂ 206 .... 318 grams.
                       American ingot iron .... 279·5 "
                       Copper .... 4·40 "
Crucible, Furnace Treatment, Degasifier, and Heat Treatment same as B₂ 199.
Weight of Casting ... 1 lb. 2 ozs.
Note.—This alloy was cast four times before obtaining a satisfactory sample.
Microphotograph... (See Plate XXXI, p. 30).

SERIES III—Continued

Samples were prepared from these alloys in the manner previously described, and mounted for exposure on the roof of Nicol building, Queens University, Kingston, Ontario. This series was mounted in the same way as Series II, and the duration of exposure was 253 days, 3 hours, beginning December 22, 1914, 1.15 p.m.—December 23, 1914, 4.30 p.m., time ending September 1, 1915, 4.15 p.m.—September 2, 1915, 7.30 p.m. Total hours, 6,075.

In addition to the samples prepared, as above described, another set of samples was prepared from the American Rolling Mill Company alloys, and given a similar heat treatment.

After having been exposed the samples were brought in, and carefully freed from rust, and final measurement made as noted in the following table:—

Table of Particulars of Corrosion Tests of Alloys Utilized in Series III.

Sample number of alloy	Approximate analysis	Diameter in cms.	Thickness in cms.	Area in square cms.	Weight in grams before exposure	Weight in grams after removal of rust	Loss of weight in grams due to corrosion	Corrosion or loss in weight in grams per sq. surface per hr. × 10⁶	Remarks
S 250 (e) (etched)¹	Fe 99·6 Co 0·25 C 0·083	2·898	0·303	16·78	15·4919	14·9704	0·5215	512	Rust light in colour and fairly hard to remove.
B 202 (a)	Fe 99·5 Co 0·25 C 0·18	2·787	0·473	16·28	22·2445	21·6166	0·6279	635	Rust fairly light in colour and easily removed.
B 202 (b)		2·755	0·741	18·17	23·5618	22·8955	0·6663	604	
C 202 (a)	Fe 99·2 Co 0·25 C 0·40	2·690	0·415	14·83	18·1622	17·7127	0·4495	499	Rust medium dark in colour and very tenacious.
C 202 (b)		2·480	0·430	12·97	16·0172	15·6251	0·3921	498	
C 202 (e) (etched)		2·181	0·709	12·24	20·1774	19·8185	0·3589	484	
S 255 (a)	Fe 99·4 Co 0·35 C 0·23	2·022	0·363	17·72	17·4452	16·9153	0·5299	493	Rust light in colour and fairly hard to remove.
S 255 (b)		2·988	0·298	17·42	17·8454	17·3428	0·5026	475	
S 255 (c)		2·781	0·320	14·90	14·7667	14·2769	0·4898	542	
S 255 (e)		2·940	0·463	17·80	24·0373	23·4600	0·5773	534	
S 265 (a)	Fe 99·3 Co 0·35 C 0·30	2·918	0·640	19·11	32·6708	31·9054	0·7654	659	Rust easily removed.
S 265 (b)		2·920	0·688	19·62	35·2712	34·5052	0·7660	643	
S 265 (c)		2·920	0·660	19·55	33·7064	32·9352	0·7712	649	
S 265 (d)		3·747	0·710	17·92	30·7998	30·1233	0·6765	623	

¹ Etched for microphotograph. See p. 30.

SERIES III—Continued.

Sample number of alloy	Approximate analysis	Diameter in cms.	Thickness in cms.	Area in square cms.	Weight in grams before exposure	Weight in grams after removal of rust	Loss of weight in grams due to corrosion	Corrosion or loss in weight in grams per sq. cm. of original surface per hr. × 10⁶	Remarks
C 255 (a)	Fe 99·2 % Co 0·35 C 0·49	2·655	0·526	15·40	21·8920	21·3656	0·5264	562	Rust difficult to remove.
C 255 (b)		2·478	0·309	11·98	11·2287	10·8213	0·4074	560	
C 255 (c)		2·247	0·656	12·52	19·6978	19·2884	0·4094	588	
B 209 (a)	Fe 99·2 % Cu 0·50 C 0·27	2·874	0·550	17·75	27·1690	26·8160	0·6546	901	Rust hard to remove.
B 209 (b)		2·620	0·652	11·50	26·9648	26·3616	0·6032	863	
B 209 (c)		2·810	0·425	14·12	19·8818	19·2750	0·6068	875	
S 252 (a)	Fe 99·1 % Cu 0·50 C 0·31	2·747	0·500	16·11	22·7013	22·2329	0·4686	479	Rust medium dark in colour, tenacious.
S 252 (b)		2·692	0·440	15·07	19·2264	18·7810	0·4454	487	
S 252 (c) (re-etched)		2·764	0·406	15·06	18·7750	18·4060	0·4690	493	
S 260 (a)	Fe 99·1 % Co 0·75 C 0·17	2·881	0·510	17·53	25·7013	25·1668	0·5345	500	Rust dark in colour and quite difficult to remove.
S 260 (b)		2·628	0·606	15·79	25·4463	24·9641	0·4822	502	
S 260 (c)		2·854	0·564	18·33	27·7345	27·2048	0·5297	472	
S 260 (d)		2·680	0·448	14·86	19·5841	19·1080	0·4761	522	
S 260 (f)		2·582	0·554	14·92	22·0557	21·5971	0·4586	502	
S 260 (g)		2·806	0·437	16·12	16·0904	15·6085	0·4819	492	
S 266 (a)	Fe 99·0 % Co 0·75 C 0·23	1·863	0·567	8·20	11·5887	11·3125	0·2762	555	Rust quite dark and difficult to remove.
S 266 (b)		2·705	0·580	16·35	25·4683	24·9446	0·4837	488	
S 266 (c)		2·733	0·542	16·34	24·1817	23·6588	0·5229	530	
S 266 (d)		2·720	0·564	16·38	25·1076	24·5862	0·5214	525	
B 198 (a)	Fe 98·6 % Co 1·0 C 0·38	2·757	0·692	16·20	31·7031	31·0973	0·6047	615	Rust difficult to remove.
B 198 (b)		2·712	0·467	15·45	20·4550	19·9428	0·5122	545	
B 198 (c)		2·703	0·547	16·04	23·9154	23·3908	0·5246	540	
S 262 (a)	Fe 98·4 % Co 1·0 C 0·62	2·632	0·470	15·88	15·4269	15·0367	0·3902	464	Rust dark in colour and hard to remove.

SERIES III—Continued.

Sample number of alloy	Approximate analysis	Diameter in cms.	Thickness in cms.	Area in square cms.	Weight in grams before exposure	Weight in grams after removal of rust	Loss of weight in grams due to corrosion	Corrosion or loss in weight in grams per sq. cm. of original surface per hr. × 10⁶	Remarks
S 261 (e)		2.805	0.405	15.11	20.8443	20.3625	0.1719	514	
B 199 (a)	Fe 97.8 Co 2.0 C 0.16	2.507	0.930	16.44	45.6640	44.9793	0.6848	686	Rust coarse-grained; easily removed.
B 199 (c)		2.780	0.672	17.94	29.9907	29.4516	0.5391	495	
B 199 (a)	Fe 97.8 Co 2.0 C 0.16	2.882	0.662	16.56	43.6886	43.0257	0.6629	659	Rust fairly difficult to remove.
B 199 (b)		2.655	0.668	16.55	28.3872	27.8941	0.4931	491	
B 199 (c) (etched)		2.780	0.635	17.60	29.6310	29.1169	0.5141	487	
B 199 (e)		1.473	0.736	8.29	10.1161	9.8990	0.2171	434	
B 200 (a)	Fe 96.8 Co 3.0 C 0.17	2.990	0.573	16.57	18.9260	18.4505	0.4755	475	Rust medium dark in colour; fairly coarse-grained.
B 200 (b)		2.880	0.530	17.77	24.0182	23.5044	0.5138	476	
B 200 (c) (etched)		2.882	0.513	17.86	25.7436	25.2580	0.4856	448	
B 200 (a)	Fe 96.6 Co 3.0 C 0.36	2.553	0.764	13.22	25.6795	25.3117	0.3678	325	Rust dark in colour and difficult to remove.
B 200 (b)		1.857	0.624	8.97	12.8802	12.6876	0.2286	420	
B 200 (e)		2.386	0.561	13.10	19.1863	18.9033	0.2830	446	Exposed 4839 hours.
S 251 (a)	Fe 99.6 Ni 0.25 C 0.057	2.905	0.744	19.92	37.8143	37.1069	0.7074	385	Rust very loose.
S 251 (b)		2.772	0.482	17.11	27.4975	26.8967	0.6010	578	
S 251 (c)		2.893	0.783	19.15	39.7612	39.0367	0.7245	502	
B 204 (a)	Fe 99.6 Ni 0.35 C 0.10	2.830	0.611	17.94	29.3386	28.7091	0.6297	578	Rust coarse-grained; easily removed.
B 204 (b)		2.568	0.552	15.10	21.8389	21.3033	0.5356	585	
S 267 (a)	Fe 99.5 Ni 0.25 C 0.25	2.740	0.464	15.51	20.7134	20.1127	0.6007	637	Rust dark in colour and difficult to remove.
S 267 (b)		2.750	0.329	14.67	14.9493	14.4270	0.5223	598	
S 267 (c)		2.740	0.575	16.64	25.8113	25.2423	0.5689	563	
S 267 (d)		2.762	0.510	16.34	23.1425	22.5593	0.5832	588	Rust dark in colour and difficult to remove.

SERIES III—Continued.

Sample number of alloy	Approximate analysis	Diameter in cms.	Thickness in cms.	Area in square cms.	Weight in grams before exposure	Weight in grams after removal of rust	Loss of weight in grams due to corrosion	Corrosion or loss in weight in grams per sq. cm. of original surface per hr. × 10⁶	Remarks
C 204 (a)	Fe 99.3 %, Ni 0.35, C 0.33	2.765	0.446	15.82	18.0501	17.5483	0.5018	522	Rust fairly loose and light colour.
C 204 (b)		2.770	0.654	18.00	29.8982	29.3469	0.5484	502	
C 204 (c)		2.830	0.667	18.62	32.3197	31.7437	0.5620	502	
S 254 (a)	Fe 99.6 %, Ni 0.35, C 0.045	2.843	0.493	16.96	23.8922	23.3215	0.5707	585	Rust coarse-grained on surface but fine grained below.
S 254 (b)		2.797	0.662	16.37	30.9796	30.3607	0.6189	624	
S 254 (c)		2.820	0.655	16.39	30.1311	29.5109	0.6202	625	
S 257 (a)	Fe 99.4 %, Ni 0.38, C 0.060	2.818	0.390	15.58	18.6770	18.1632	0.5138	543	Rust medium in colour and quite tenacious.
S 257 (b)		2.820	0.421	16.18	20.1592	19.6474	0.5218	534	
S 258 (a)	Fe 99.4 %, Ni 0.50, C 0.072	2.905	0.362	16.30	18.5816	18.0294	0.4522	454	Rust very dark in colour and removed with great difficulty.
S 258 (c)		2.885	0.544	17.92	22.7468	22.2670	0.4798	442	
S 268 (a)	Fe 99.3 %, Ni 0.50, C 0.21	2.479	0.579	14.04	21.4494	21.0221	0.4173	489	Rust medium dark in colour and quite fine-grained.
S 268 (b)		2.449	0.585	10.61	14.8962	14.4988	0.4074	476	
S 268 (c)		2.780	0.570	17.09	26.5809	26.4139	0.1670	537	Exposed 4896 hours.
S 259 (a)	Fe 99.4 %, Ni 0.78, C 0.067	2.905	0.532	18.03	27.2486	26.6656	0.5839	533	Rust fairly loose and light in colour.
S 259 (b)		2.865	0.425	16.64	21.0132	20.4614	0.5518	537	
S 259 (c)		2.949	0.544	16.81	18.0560	17.4776	0.5284	565	
S 259 (d)		2.780	0.464	16.13	21.6476	21.1233	0.5343	533	
S 259 (e) oil chest		2.901	0.596	18.56	30.4007	29.8090	0.5917	525	
S 263 (e)	Fe 98.9 %, Ni 1.0, C 0.065	2.949	0.357	16.56	17.5516	17.0942	0.4574	485	Rust dark and very hard to remove.
B₂ 195 (a)	Fe 98.9 %, Ni 1.0, C 0.080	2.875	0.539	16.16	16.9031	16.4033	0.4998	509	Rust dark in colour and hard to remove.
B₂ 195 (b)		2.802	0.637	17.67	30.3503	29.8193	0.5310	495	
B₂ 195 (c)		2.882	0.512	17.63	25.6145	25.0789	0.5356	500	

SERIES III—Continued.

	Dimensions of Disc before Exposure					*Weight of Disc*			
Sample number of alloy	Approximate analysis	Diameter in cms.	Thickness in cms.	Area in square cms.	Weight in grams before exposure	Weight in grams after removal of rust	Loss of weight in grams due to corrosion	Corrosion or loss in weight in grams per sq. cm. of original surface per hr. × 10⁸	Remarks
B 195 (b)	Fe 98.8 % Ni 1.0 C 0.14	2.710	0.449	14.84	19.9142	19.4903	0.4239	470	Rust very dark in colour and difficult to remove.
B 195 (c)		2.820	0.585	17.52	28.2365	27.7448	0.4917	462	
S 269 (a)	Fe 97.9 % Ni 2.0 C 0.085	2.920	0.345	16.51	17.6610	17.1717	0.4893	488	Rust dark in colour, coarse-grained and hard to remove.
S 269 (b)		2.880	0.574	18.12	27.4780	26.9605	0.5175	470	
S 269 (c)		2.924	0.583	18.72	29.1942	28.7006	0.4936	538	Exposed 4896 hours.
B 196 (a)	Fe 97.8 % Ni 2.0 C 0.23	2.865	0.638	18.57	31.7648	31.2547	0.5101	452	Rust dark, fine-grained and quite tenacious.
B 196 (b)		2.810	0.720	18.69	34.5114	34.0049	0.5065	447	
B 196 (c)		2.696	0.645	16.30	28.2857	27.8783	0.4074	495	
B₂ 197 (a)	Fe 96.8 % Ni 3.0 C 0.13	2.938	0.361	16.78	18.3561	17.9187	0.4374	429	Rust very dark and fine-grained quite tenacious
B₂ 197 (b)		2.982	0.232	15.77	12.1361	11.7085	0.4276	446	
B₂ 197 (c)		2.818	0.224	14.39	10.6776	10.2948	0.3830	438	
B 197 (a)	Fe 96.7 % Ni 3.0 C 0.21	2.733	0.621	16.94	28.0926	27.6833	0.4093	398	Rust very dark in colour.
B 197 (b)		2.740	0.436	15.45	19.4972	19.1351	0.3621	386	
B 197 (c)		2.679	0.568	15.93	24.5124	24.1458	0.3666	379	
B₂ 205 (a)	Fe 99.7 % Cu 0.25 C 0.045	2.520	0.497	13.94	17.0521	16.4935	0.5586	663	Rust very loose and light in colour.
B₂ 205 (b)		2.768	0.464	16.02	21.2016	20.5370	0.6646	683	
B₂ 205 (c)		2.757	0.283	14.53	12.9512	12.3301	0.6211	712	
B₂ 205 (d)		2.790	0.584	17.22	27.5268	26.7933	0.7335	701	
B 205 (a)	Fe 99.6 % Cu 0.25 C 0.19	2.790	0.653	17.85	30.6414	30.1192	0.5222	482	Rust fairly dark in colour and tenacious.
B 205 (b)		2.800	0.663	18.10	31.2587	30.7331	0.5256	478	
B 205 (c)		2.783	0.565	17.05	26.4556	26.0039	0.4517	541	Exposed 4896 hours.
B₂ 206 (a)	Fe 99.3 % Cu 0.50 C 0.17	2.777	0.556	16.85	25.8509	25.4009	0.4500	440	Rust light in colour and difficult to remove.
B₂ 206 (b)		2.665	0.674	16.73	29.0202	28.5702	0.4500	443	

SERIES III—Continued.

Sample number of alloy	Approximate analysis	Diameter in cms.	Thickness in cms.	Area in square cms.	Weight in grams before exposure	Weight in grams after removal of rust	Loss of weight in grams due to corrosion	Corrosion or loss in weights in grams per sq. cm. of original surface per hr. × 10⁶	Remarks
B 206 (e)		2·805	0·886	17·44	27·8550	27·3661	0·4889	443	
B 206 (a)	Fe 99·4 % Cu 0·50 C 0·19	2·835	0·875	17·59	29·0498	28·5345	0·5153	482	Rust light in colour and loose.
B 206 (b)		2·833	0·734	18·94	38·4891	37·9390	0·5501	478	
B 206 (c)		2·940	0·997	19·92	36·1260	35·5740	0·5520	456	
B 208 (a)	Fe 99·0 % Cu 0·75 C 0·18	2·570	0·824	14·51	22·7248	22·2673	0·4575	519	Rust dark in colour and fairly tenacious.
B 208 (b)		2·620	0·272	13·81	15·4644	15·0484	0·4160	497	
B 208 (c)		2·960	0·537	18·54	27·8213	27·2679	0·5534	492	
34204 (a)	American ingot iron 99·6 % Co 0·35 C 0·01	3·238	0·734	23·84	46·9786	46·2114	0·7672	638	Rust light in colour and fairly tenacious. Exposed 5043 hours.
34204 (b)		2·959	0·823	21·34	44·0585	43·3660	0·6925	643	Rust light in colour and fairly tenacious. Exposed 5043 hours.
34204 (c)		3·017	0·816	21·95	45·4052	44·6974	0·7078	638	
34204 (d)		3·140	0·814	23·46	48·9962	48·2394	0·7568	638	
34196 (a)	American ingot iron 99·3 % Co 0·60 C 0·01	3·008	1·004	23·60	55·6110	54·6935	0·9175	640	Rust light in colour, scaly.
34196 (b)		2·930	1·025	23·82	54·8524	54·0112	0·8412	582	
34196 (c)		3·105	0·975	24·84	55·0917	54·0661	0·9256	613	
34196 (d)		2·993	0·999	23·33	54·8672	53·8681	0·9991	706	
34196 (f)		3·007	0·998	23·57	55·3730	54·4791	0·8939	624	
34185 (a)	American ingot iron 98·8 % Co 1·18 C 0·01	3·138	1·006	25·27	60·5641	59·7725	0·7916	522	Rust medium in colour and fairly difficult to remove. Exposed 6018 hours; found on roof.
34185 (b)		3·180	0·977	25·85	60·4420	59·6150	0·8270	533	
34185 (c)		3·172	1·000	25·70	61·6745	60·8444	0·8301	532	
34185 (d)		3·185	1·011	25·47	62·9224	61·9945	0·8279	535	

¹ These samples were mounted February 4, 1915, at 2.35 p.m.

Following, are a series of microphotographs of the alloys described under Series III:—

PLATE III.

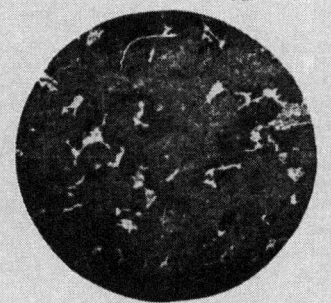

Alloy B 202
Fe ... 99·5%
Co ... 0·25
C ... 0·18

PLATE IV.

Alloy B₃ 199
Fe ... 97·8%
Co ... 2·0
C ... 0·16

PLATE V.

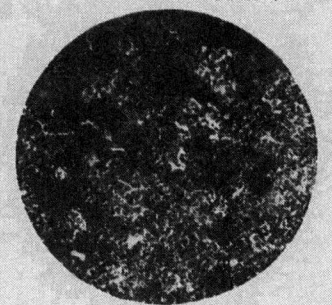

Alloy S 250
Co   0·25%
C   0·083
Fe   99·6

PLATE VI.

Alloy S 262
Fe   98·3%
Co   1·0
C   0·62

PLATE VII.

Alloy S 255
Fe ... 99.4%
Co ... 0.35
C ... 0.21

PLATE VIII.

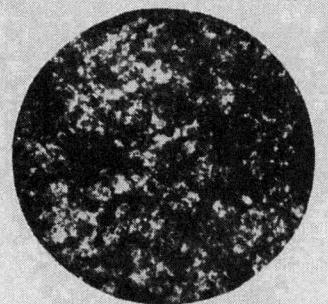

Alloy B 209
Fe ... 99.2%
Co ... 0.5
C ... 0.27

PLATE IX.

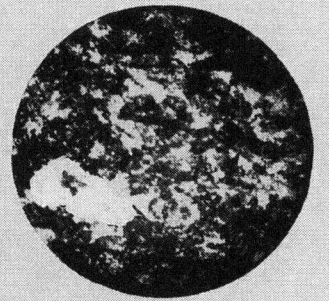

Alloy S 252
Fe ... 99.1%
Co ... 0.50
C ... 0.31

PLATE X.

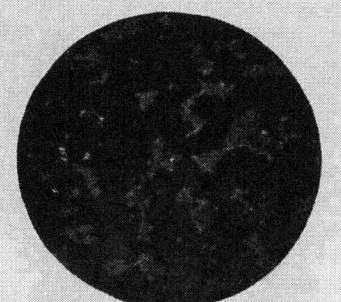

Alloy S 260
Fe ... 99.1%
Co ... 0.75
C ... 0.17

PLATE XI.

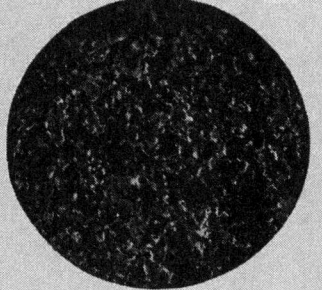

Alloy B 198
Fe ... 98.6%
Co ... 1.0
C ... 0.38

PLATE XII.

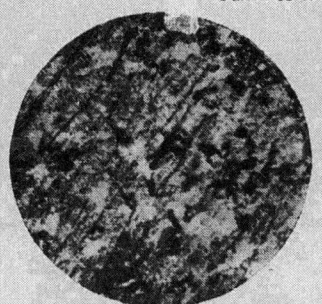

Alloy B₂ 200
Fe ... 96.8%
Co ... 3.0
C ... 0.17

PLATE XIII.

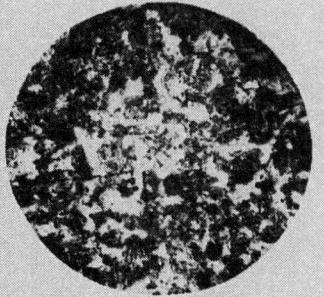

Alloy C 202
Fe .... 99·2%
Co .... 0·25
C .... 0·49

PLATE XIV.

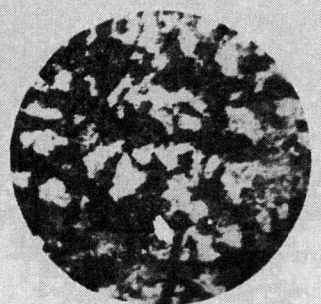

Alloy C 255
Fe .... 99·2%
Co .... 0·35
C .... 0·49

PLATE XV.

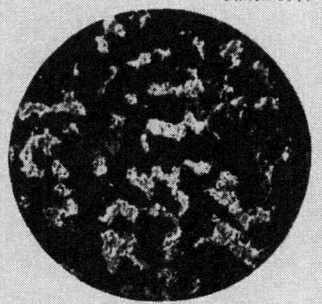

Alloy B 199
Fe ... 97·5%
Co ... 2·0
C ... 0·46

PLATE XVI.

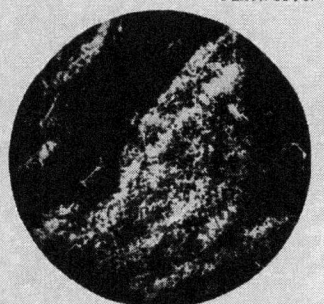

Alloy S 251
Fe ... 99·6%
Ni ... 0·25
C ... 0·057

PLATE XVII.

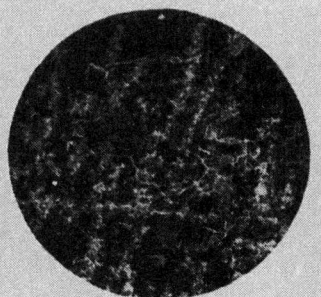

Alloy S 254
Fe   99·6
Ni   0·35
C   0·045

PLATE XVIII.

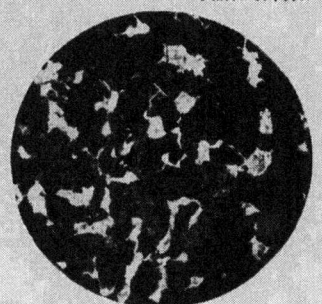

Alloy S 258
Fe   99·4
Ni   0·50
C   0·072

PLATE XIX.

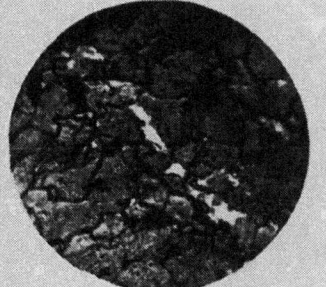

Alloy S 259
Fe   99.2%
Ni   0.75
C   0.067

PLATE XX.

Alloy B. 195
Fe   98.9%
Ni   1.0
C   0.089

PLATE XXI.

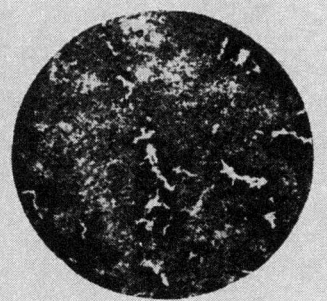

Alloy S 263
Fe ... 98.9%
Ni ... 1.0
C ... 0.065

PLATE XXII.

Alloy S 269
Fe ... 97.9%
Ni ... 2.0
C ... 0.085

PLATE XXIII.

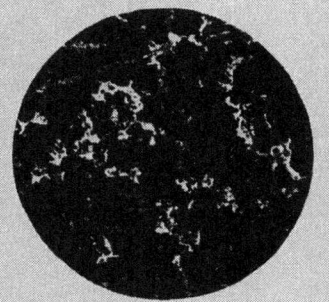

Alloy B 195
Fe   98.8%
Ni   1.0
C   0.21

PLATE XXIV.

Alloy B 196
Fe   97.8%
Ni   2.0
C   0.23

PLATE XXV.

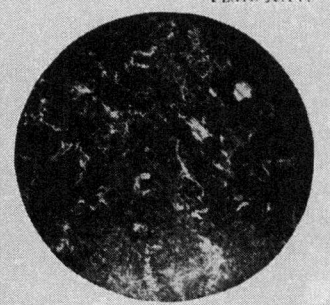

Alloy B 197
Fe   96·8%
Ni   3·0
C   0·14

PLATE XXVI.

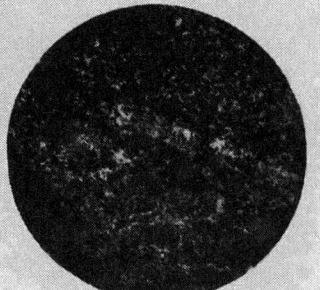

Alloy B 197
Fe   96·7%
Ni   3·0
C   0·21

PLATE XXVII.

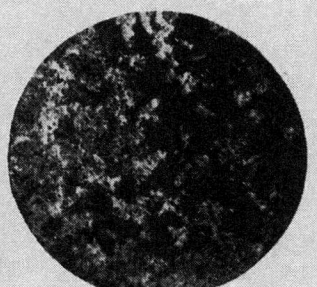

Alloy C 201
Fe   99.5%
Ni   0.25
C   0.13

PLATE XXVIII.

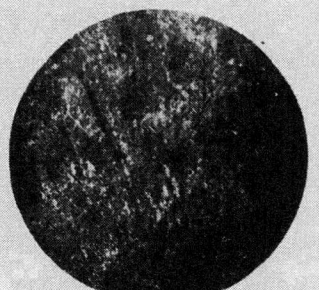

Alloy B₁ 205
Fe   99.7%
Cu   0.25
C   0.045

PLATE XXIX.

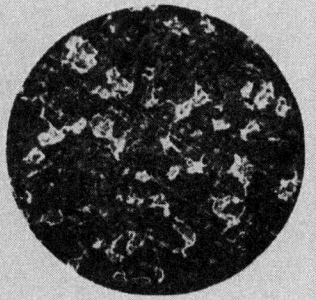

Alloy B₂ 206
Fe   99.3%
Cu   0.50
C   0.17

PLATE XXX.

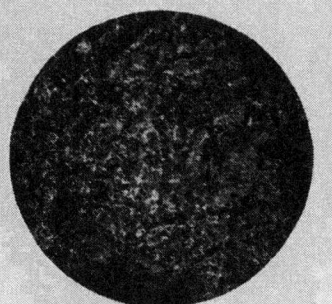

Alloy B 206
Fe   99.3%
Cu   0.50
C   0.19

PLATE XXXI.

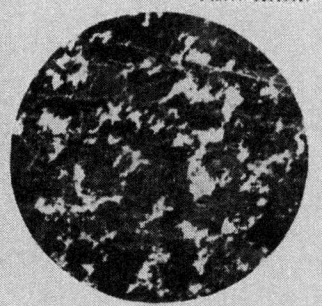

Alloy B 208
Fe 99·0
Cu 1·0
C 0·18

SERIES IV

Accelerated Corrosion Tests.

A few accelerated corrosion tests were made on some of these alloys, the results of which may be instructive, although not conclusive. The authors do not believe that much reliance can be placed upon conclusions as to what would occur under service conditions in the atmosphere, which are drawn from accelerated tests. These measurements are included with the rest for completeness.

Accelerated corrosion tests were made on the five heats of ingot iron alloys submitted by the American Rolling Mill Company as follows:

No. 34175	S.	0·026
	P.	0·009
	C.	0·010
	Mn.	0·022
	Cu.	0·016
No. 34185	S.	0·034
	P.	0·006
	C.	0·015
	Mn.	0·017
	Cu.	0·028
	Co.	1·18
No. 34196	S.	0·040
	P.	0·008
	C.	0·010
	Mn.	0·020
	Cu.	0·024
	Co.	0·60
No. 34204	S.	0·022
	P.	0·009
	C.	0·010
	Mn.	0·020
	Cu.	0·020
	Co.	0·35
No. 44009	S.	0·025
	P.	0·008
	C.	0·010
	Mn.	0·015
	Cu.	0·24
	Ni.	0·75

Method of Making Accelerated Corrosion Tests

These tests were made either by (a) immersing the samples in the form of spheres in a dilute sulphuric acid for a period of 1 hour, and noting the loss in weight, or, (b) submitting them to the intermittent action of dilute sulphuric acid and the atmosphere. The acid used for the tests was 20% H_2SO_4.

SERIES IV—Continued.

Samples were used throughout in the form of spheres of approximately 7 sq. cms. surface.

The intermittent corrosion tests were made by immersing the samples in the acid in one compartment of a wooden box, so arranged that the samples were covered and uncovered automatically by the tilting of this box over a knife edge pivot by the action of water from the water tap. The balls were held in place at one end of the box by pieces of glass rod.

The apparatus used for making these corrosion tests consisted of a wooden box about 18″ long by 12″ wide, with sides about 3″ high. This box was divided into two lengthwise compartments, one larger than the other, and the larger one being still further divided into two equal compartments, this division being across the narrow way of the box directly in the middle, the dividing board extending well above the edge of the box. Across the outside of the bottom of the box, in the middle, a wooden knife edge was fastened.

The operation was somewhat as follows: the samples were placed at one end of the undivided lengthwise compartment, supported and held in place by glass rods, and covered with the corroding solution. This unbalanced the box so that it inclined to one side, keeping the samples under the solution until the box was tilted to the opposite side. This tilting was accomplished by opening the water tap above the corrosion box, which allowed the stream of water to run into the high side of the large divided compartment, this being the end opposite to the one in which the samples were placed. As soon as this compartment was sufficiently full of water to overbalance the weight of the corroding solution and the samples, the box tilted, causing the solution to flow away from the samples, and at the same time directing a stream of water from the tap, by means of the extended dividing board, into the other compartment, which again tended to overbalance the box and cause it to assume its original position.

In the end of each of the water compartments, just above the water line, or the line to which it was necessary for the water to rise to overbalance the box, a hole was made which allowed the water to run out after it had been overbalanced. On account of the extra weight of the samples and glass rods it was necessary to have this water line fairly high, thus requiring an automatic syphon, consisting of a bent glass tube inserted into the hole, and touching the bottom of the box, to drain completely. The action of the box was made eccentric, by attaching to one side of the water-diverting board a small trough, which carried away into the sink most of the water flowing from the tap on that side. The compartment on this side of the box was filled by the water, which dropped through holes made in the trough for this purpose. The stream from the tap and these holes were so adjusted that a complete period of the box was about one-half hour.

Accelerated Corrosion Measurements

On May 21, 1914, the average dimension of the spheres before corrosion was found to be as follows:

Sample Number	Diameter in cms.	Surface in sq. cms.	Weight in grams before corrosion	Final weight in grams	Loss in weight in grams	Loss in weight per sq. cm. of exposed surface per hour
34125	1.2656	7.952	8.3438	8.3275	0.0158	0.00216
34185	1.2436	7.808	7.8775	7.8608	0.0167	0.00238
34196	1.1811	7.421	6.7720	6.7584	0.0136	0.00204
34204	1.1739	7.376	6.6582	6.6440	0.0142	0.00214
44009	1.1928	6.867	5.4585	5.4513	0.0072	0.00117

SERIES IV—Continued

During the course of these experiments, the 5 spheres were each immersed in the acid for 54 minutes, and exposed to the air within the room of the laboratory for 5 hours and 5 minutes. These periods of immersion and exposure were each divided into 12 approximately equal intervals, the twelve 4½-minute immersions, with the corresponding twelve exposures, constituting the corrosion test.

In columns 5 and 6 of the above table are given the final weight in grams, after the corrosion test and corresponding loss in weight.

A check was run of the above corrosion test as indicated in the following table:—

(May 26, 1914.)

Sample Number	Diameter in cms.	Surface in sq. cms.	Weight in grams before corrosion	Final weight in grams	Loss in weight in grams	Loss in weight per sq. cm. of exposed surface per hour
34175		7.984	8.3199	8.3033	0.0166	0.00252
34185		7.908	7.8532	7.8352	0.0180	0.00256
34196		7.421	6.7540	6.7386	0.0144	0.00216
34204		7.376	6.6374	6.6222	0.0152	0.00229
44009		6.867	5.3484	5.3404	0.0079	0.00246

It will be seen from the above tables, that the two sets of measurements agree very well among themselves, and that the order of passivity of the alloys is as follows:—

44009
34196
34204
34175
34185

For comparison of the above, the standard sulphuric acid accelerated corrosion test was run with the modified one just described. These tests were made by immersing the five samples above described in 20% sulphuric acid continuously for 54 minutes. The following table gives the results:

Sample Number	Weight in grams before immersion	Weight in grams after immersion	Loss in weight in grams	Loss in weight in grams per sq. cm. of exposed surface per hour
34175	8.4274	8.4235	0.0048	0.00053
34185	7.8608	7.8569	0.0039	0.00053
34196	6.7584	6.7557	0.0027	0.00040
34204	6.6440	6.6406	0.0034	0.00051
44009	5.3514	5.3498	0.0016	0.00026

SERIES IV—*Continued.*

A check run on this corrosion test, identical with the above, was made with the following results:—

Sample Number	Weight in grams before immersion	Weight in grams after immersion	Loss in weight in grams	Loss in weight in grams per sq. cm. of exposed surface per hour
34175	8.4245	8.4199	0.0046	0.00042
34185	7.8569	7.8532	0.0037	0.00053
34196	6.7557	6.7530	0.0027	0.00040
34204	6.6406	6.6374	0.0032	0.00050
44009	5.3498	5.3483	0.0015	0.00024

These two tests agree with one another, and show the order of passivity of the alloys to be as follows:—

44009
34196
34204
34175
34185

Conclusion

If these accelerated corrosion tests could be relied upon as accurately reproducing atmospheric conditions, it would be clear that the addition of monel metal to American ingot iron, to the extent of about 1%, produces a more non-corrosive alloy for sheet roofing materials than the addition of similar small percentages of cobalt.

This type of corrosion test shows the alloy containing 0.60% Co to be more non-corrosive than either the one containing more (1.18% Co) or the one containing less (0.35% Co).

Diagram I
Series II.

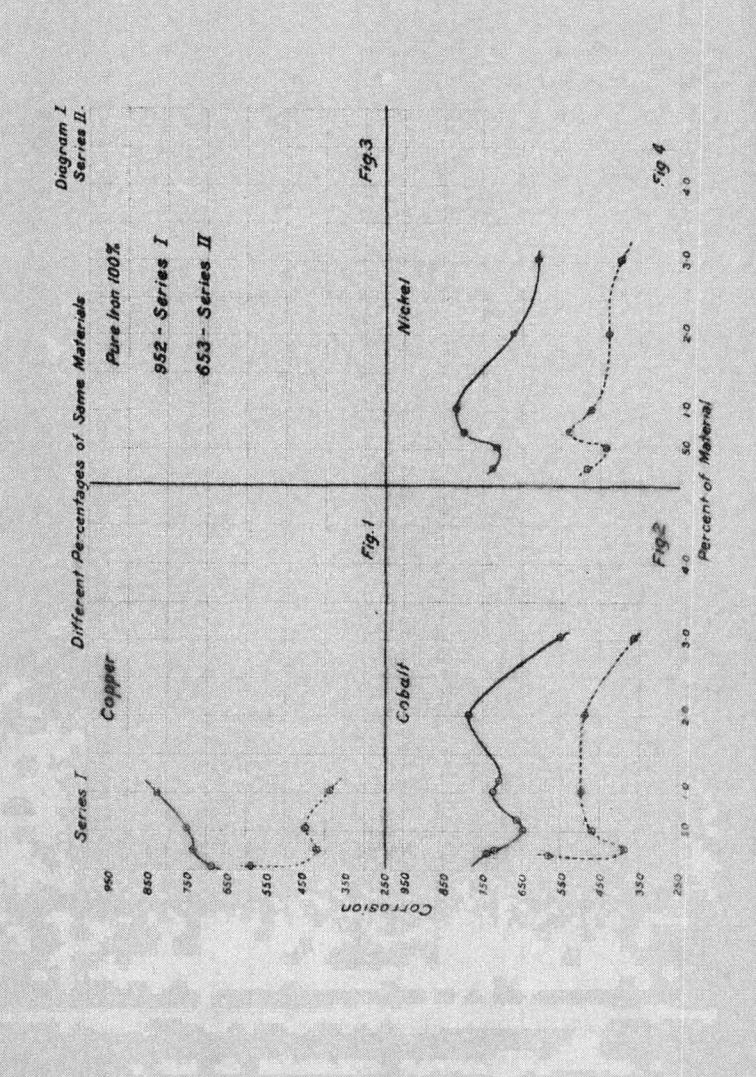

Different Percentages of Same Materials

Pure Iron 100%

952 - Series I
653 - Series II

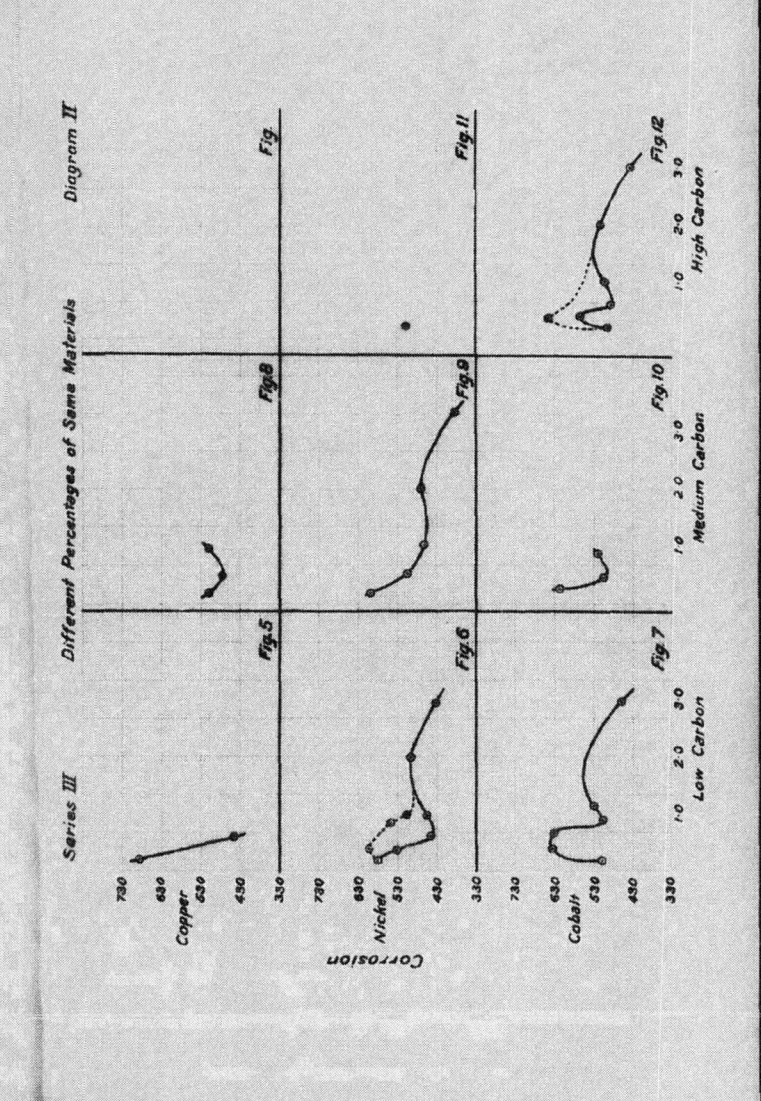

Series III   Different Percentages of Same Materials   Diagram II

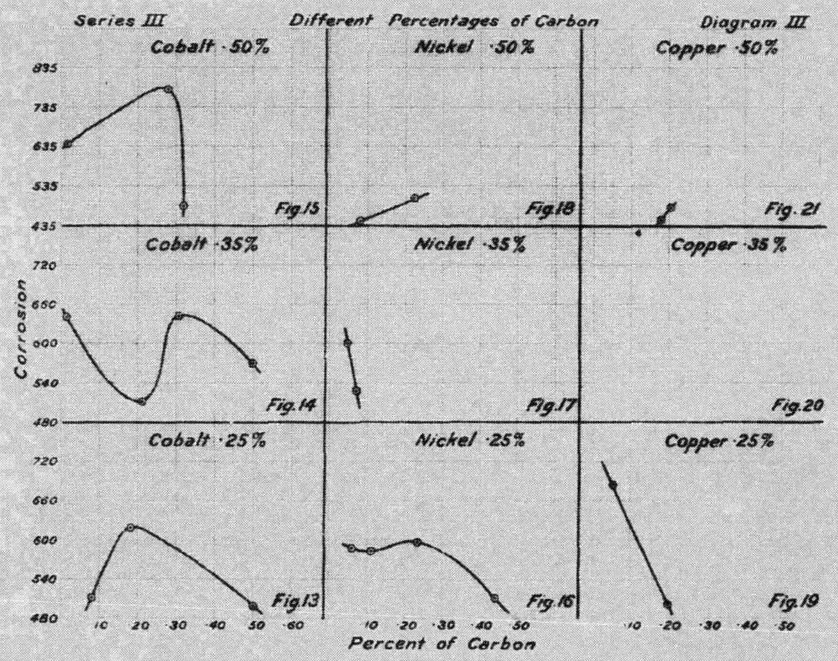

Series III
Different Percentages of Carbon
Diagram III
Cobalt ·50%
Nickel ·50%
Copper ·50%
835
785
635
535
435
Fig.15
Fig.18
Fig. 21
Cobalt ·35%
Nickel ·35%
Copper ·35%
720
660
600
540
480
Fig.14
Fig.17
Fig.20
Cobalt ·25%
Nickel ·25%
Copper ·25%
720
660
600
540
480
Fig.13
Fig.16
Fig.19
Corrosion
·10 ·20 ·30 ·40 ·50 ·60
·10 ·20 ·30 ·40 ·50
·10 ·20 ·30 ·40 ·50
Percent of Carbon

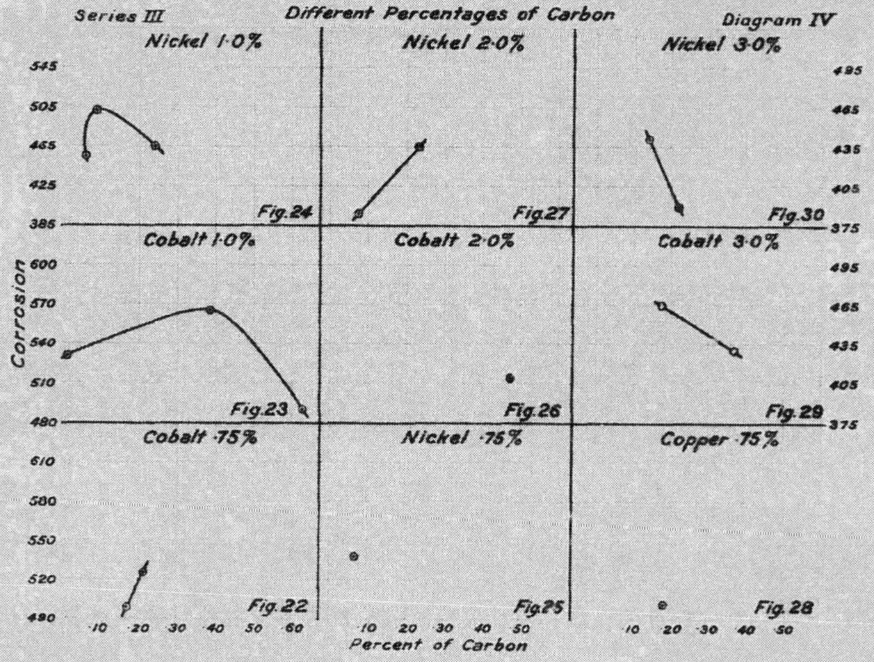

Series III
Different Percentages of Carbon
Diagram IV
Nickel 1·0%
Nickel 2·0%
Nickel 3·0%
Fig. 24
Fig. 27
Fig. 30
Cobalt 1·0%
Cobalt 2·0%
Cobalt 3·0%
Fig. 23
Fig. 26
Fig. 29
Cobalt ·75%
Nickel ·75%
Copper ·75%
Fig. 22
Fig. 25
Fig. 28
Corrosion
545
505
465
425
385
600
570
540
510
480
610
580
550
520
490
495
465
435
405
375
495
465
435
405
375
·10 ·20 ·30 ·40 ·50 ·60
·10 ·20 ·30 ·40 ·50
·10 ·20 ·30 ·40 ·50
Percent of Carbon

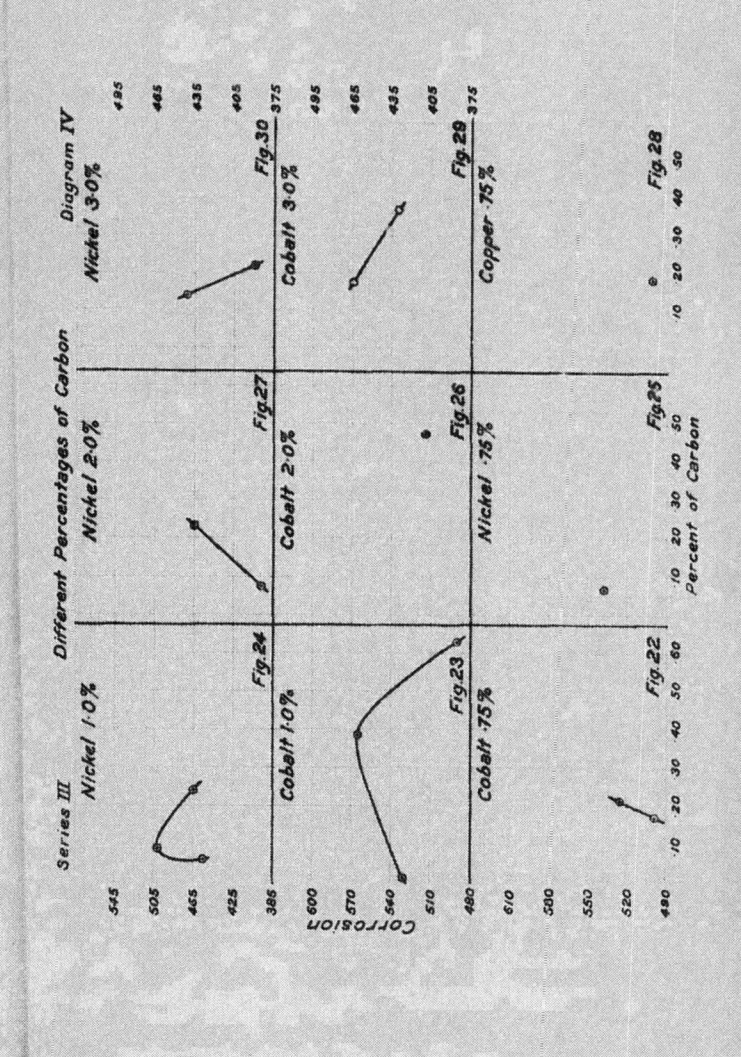

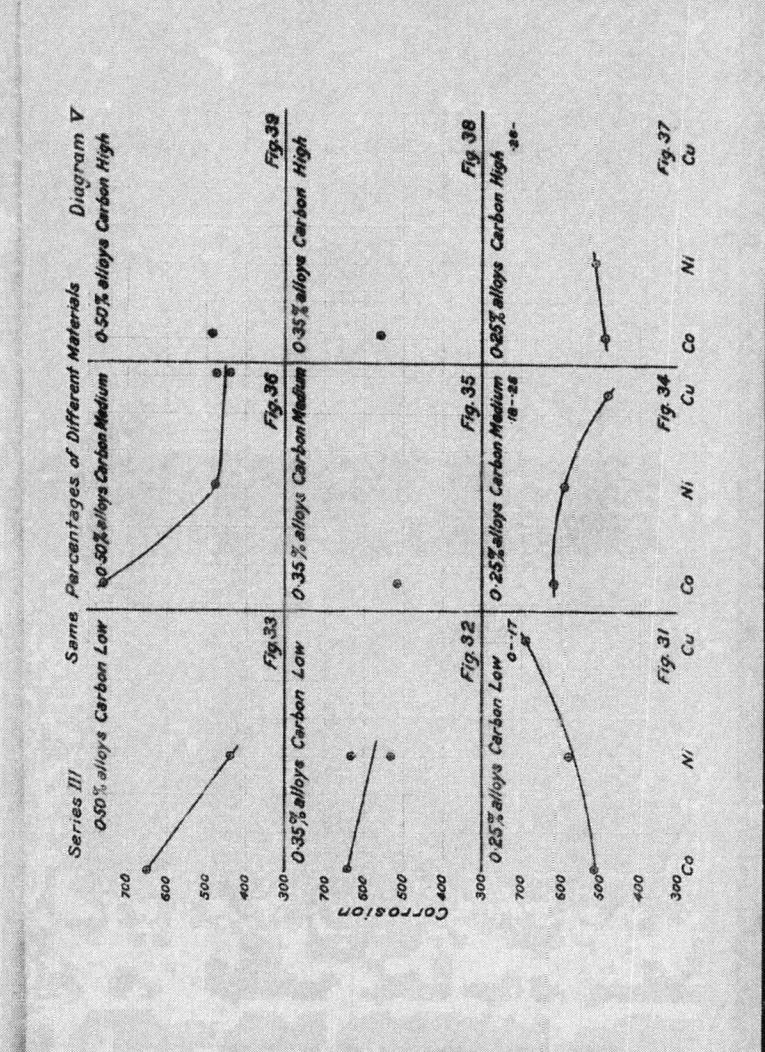

Series III Same Percentages of Different Materials Diagram V

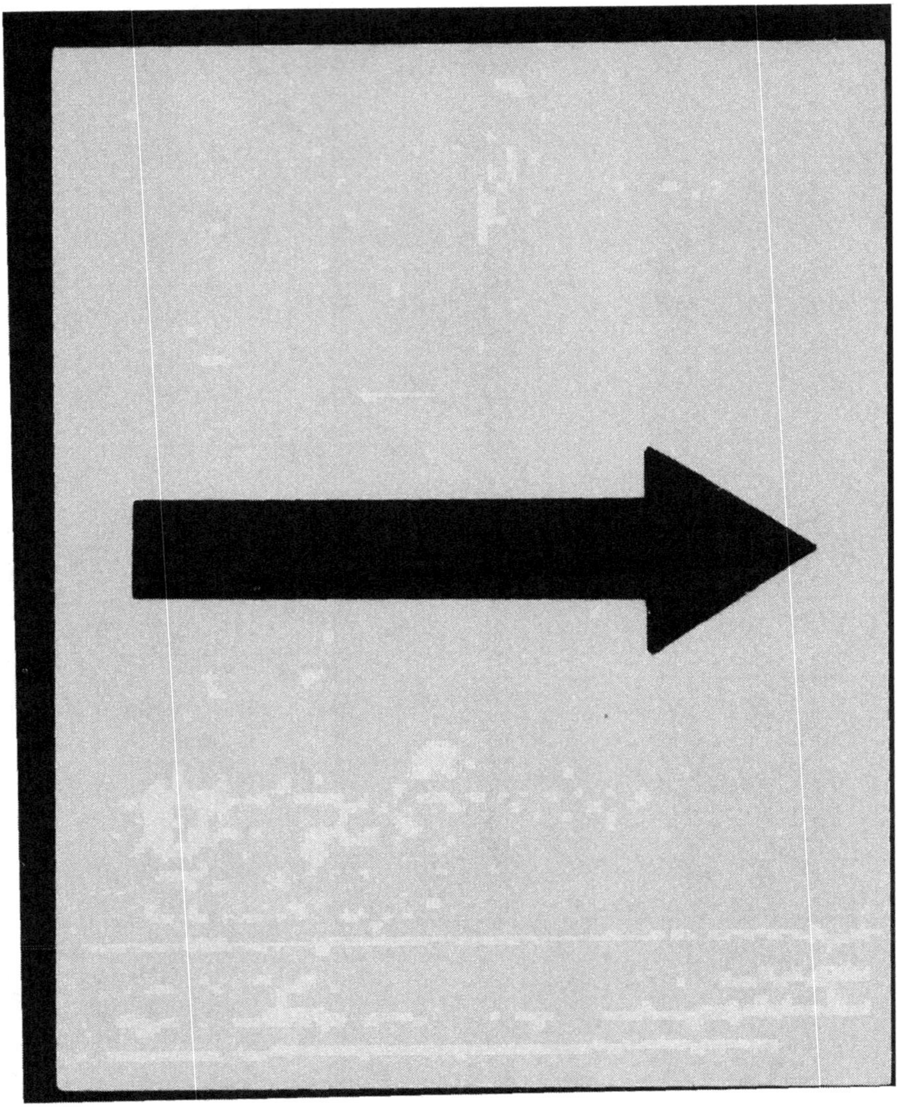

MICROCOPY RESOLUTION TEST CHART

(ANSI and ISO TEST CHART No. 2)

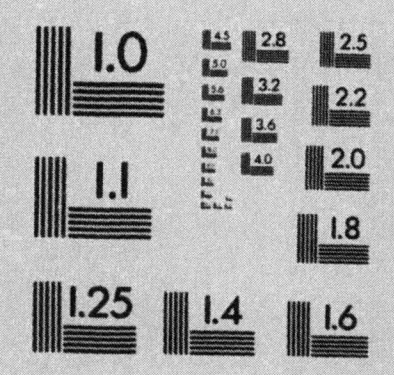

APPLIED IMAGE Inc
1653 East Main Street
Rochester, New York 14609 USA
(716) 482-0300 – Phone
(716) 288-5989 – Fax

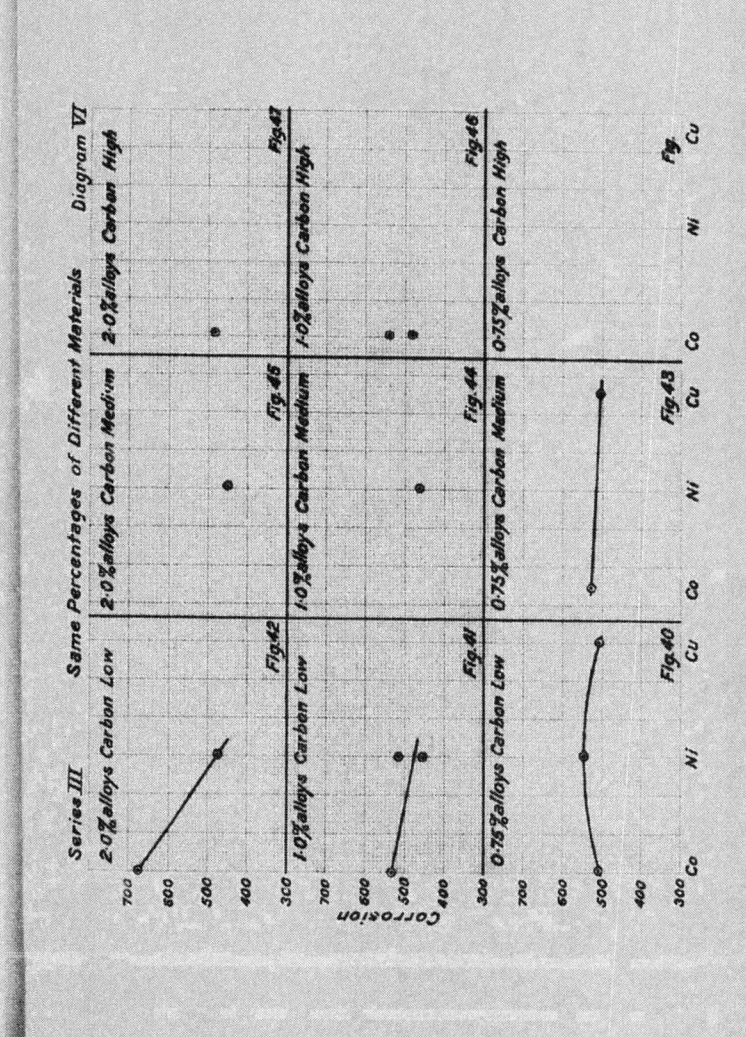

Series III Same Percentages of Different Materials Diagram VI

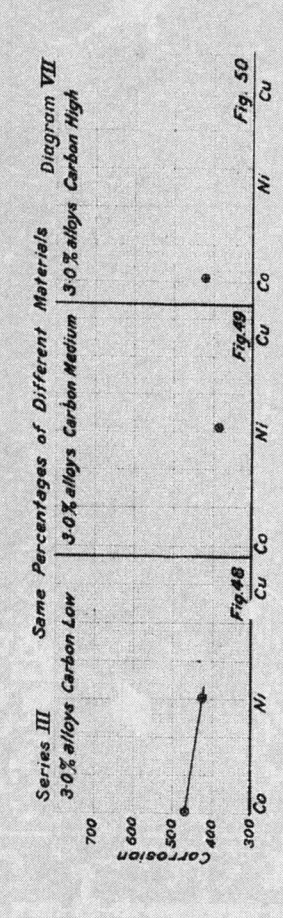

Series III Same Percentages of Different Materials Diagram VII
3.0% alloys Carbon Low | 3.0% alloys Carbon Medium | 3.0% alloys Carbon High

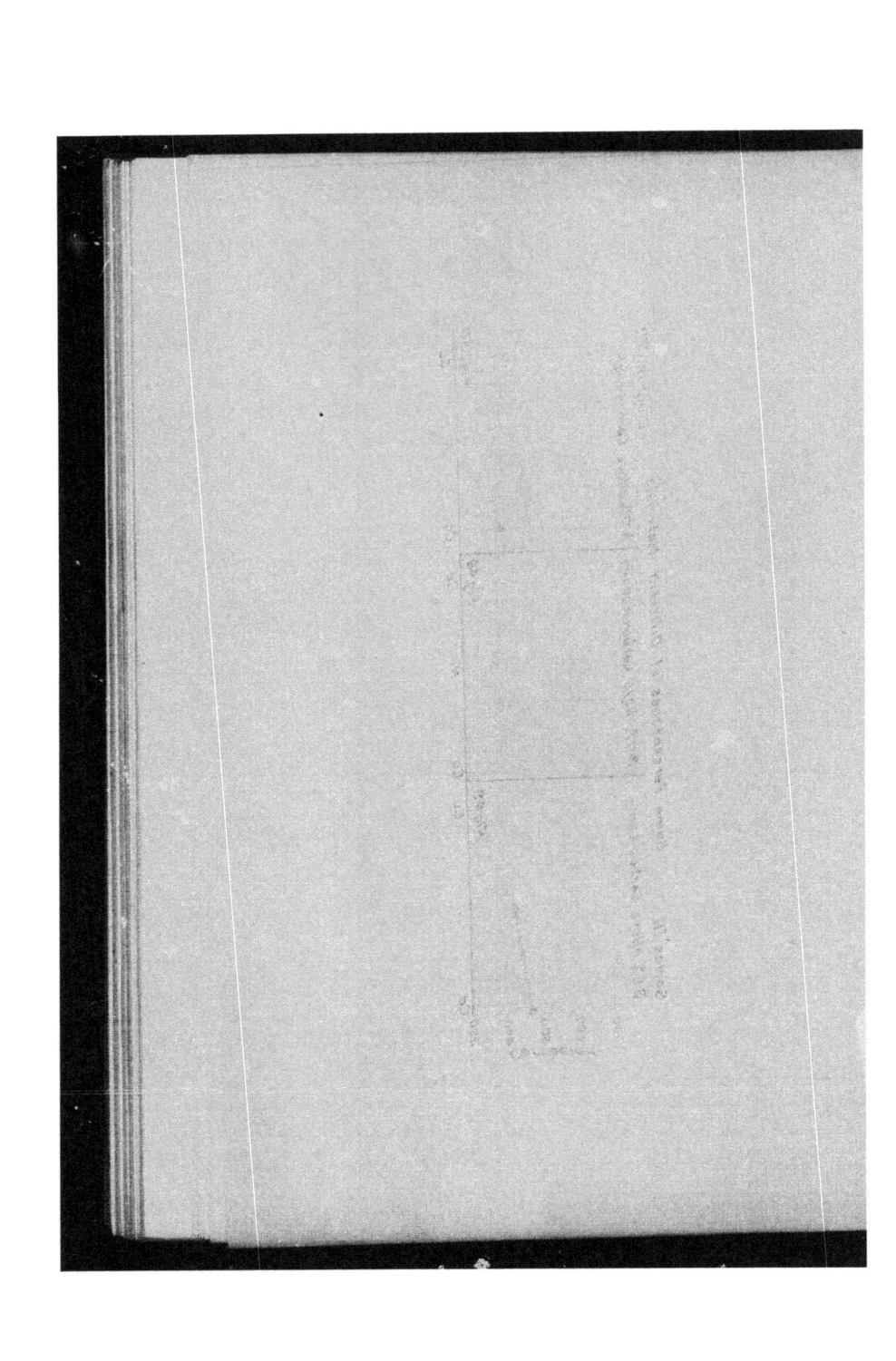

CORROSION TESTS, AMERICAN INGOT IRON ALLOYS IN THE FORM OF SHEET-ROOFING MATERIALS.

In addition to the numerous corrosion experiments on discs of the alloy above mentioned, there were atmospheric corrosion tests made on sheets of these alloys under service conditions.

Pure cobalt of the analysis shown on page 5 was sent from this laboratory to the American Rolling Mill Company, Middletown, Ohio, early in 1913. It was arranged to make up charges in their four-ton experimental furnace, such as could be rolled into full-sized roofing sheets to be subjected to the usual corrosion tests. Three such cobalt alloy sheets were sent to this laboratory, together with a run similarly prepared, using monel metal in place of cobalt and a sheet of standard American ingot iron for comparison. These sheets were 30" by 96" in dimensions and No. 26 gauge (0.0188" in thickness). All of these sheets were received in duplicate.

Analyses

Following are the analyses of the above mentioned sheets.

Analyses of Sheet Roofing Materials

American Ingot Iron 34175	1% Co Alloy 34185	0.60% Co Alloy 34196	0.35% Co Alloy 34204	1% Monel Metal 43009
S 0.026	S 0.014	S 0.040	S 0.022	S 0.025
P 0.009	P 0.006	P 0.008	P 0.009	P 0.008
C 0.010	C 0.015	C 0.010	C 0.010	C 0.010
Mn 0.022	Mn 0.017	Mn 0.020	Mn 0.020	Mn 0.015
Cu 0.016	Cu 0.028	Cu 0.024	Cu 0.020	Cu 0.24
	Co 1.18	Co 0.60	Co 0.35	Ni 0.75

These five sheets were mounted side by side on a wooden frame built to support them. They were exposed in a plane making 60 degrees with a horizontal. They have been corroding since March 18, 1914, and since that time photographic and visual observations have been made at regular intervals.

It will take at least another year for these sheets to corrode through to destruction, before which time no final conclusions can be drawn. Up to this time it would appear, however, that the loss of weight of the 1% cobalt alloy, and of the 0.60% cobalt alloy, was less than that of the others. Both of these have formed a very hard, dense, tenacious, protective coating.

While there may be some doubt as to which is the least corroded sheet of the three; the 1% cobalt alloy; the 0.60% cobalt alloy; or the 1% monel metal alloy, there can be no doubt as to the fact that all of these three are far superior in non-corrosive action to the sheet of pure American ingot iron, or to the sheet containing 0.35% cobalt alloy. The pure American ingot iron sheet is the most corroded to date.

Following are diagrams i to vii, containing Figs. 1 to 50, these being graphs for the purpose of visualizing the foregoing data. From the data and graphs, the following general conclusions may be drawn.

GENERAL CONCLUSIONS

1. The corrosion or loss of weight in grams per square centimetre of original surface per hour, is a function of the length of exposure, being less for the longer exposures. This is true, in part at least, because of the property of these alloys to form a self-protecting layer or coating.

2. The graphs representing Series I and II show remarkable similarity of form. These series were for independent corrosions of the same set of samples. Therefore, such irregularities as appear in general, are probably due not to uncertainties of measurement, but to lack of control of the physical structure of the alloys in preparation.

3. The alloys formed by the addition of small percentages of copper, nickel, and cobalt (from 0·25% to 3·0%) to American ingot iron, are more resistant to atmospheric corrosion than the pure American ingot iron, from which the alloys were prepared.

4. Considering the data for alloys formed by adding various amounts of cobalt (from 0·25% to 3·0%) to American ingot iron, with very little, if any, carbon content, it is apparent that the corrosion is not a simple function of the percentage of cobalt content. In general, the corrosion of the alloys formed by the addition of 3% of cobalt to American ingot iron, is about 75% as great as that of the alloys formed by the addition of 0·5% of cobalt.

5. Alloys formed by the addition of 0·25% to 3·0% cobalt to American ingot iron, with very little, if any, carbon content, are corroded in the atmosphere to an extent varying between 50% and 75% of that of the pure American ingot iron, from which the alloys were prepared.

6. Conclusions 4 and 5 are, approximately, as true for the corresponding nickel alloys as for the cobalt alloys. There seems to be very little choice between the use of nickel and cobalt to form alloys with American ingot iron containing between 0·25% and 3·0% of the added metal for the prevention of corrosion. This is true so far as the disc tests show, but with the exception noted in conclusion 7.

7. As corrosion progresses, all the alloys prepared form self-protective coatings of oxides. It is noticeable throughout, that the oxides formed by the cobalt are darker, denser   re tenacious than those formed by the other alloys.

8. The effect of preventin   of the protective coating mentioned in conclusion 7, does not s   worked greatly to the advantage of the cobalt alloys, in spite of   satisfactory appearance, in the length of time that the above experiments were allowed to run.

9. In order to finally conclude the possible ultimate advantages of the cobalt alloy protective coating, as compared with the other sheets, all of the alloys should be allowed to corrode to destruction. The results of such tests, as discussed in the text above, will be published later.

10. The addition of copper to American ingot iron to an extent between 0·25% and 0·75%, seems to be conducive to reducing the corrosion of American ingot iron under atmospheric conditions. It is difficult to say whether or not the addition of copper in these amounts has a greater or lesser effect than the corresponding amounts of nickel or cobalt. Additional experiments will be required to determine these facts, but there can be little doubt that, the addition of copper, as above reported, diminishes the corrosion of the pure American ingot iron.

11. The amount of corrosion varies with the percentage of carbon in the alloy, as would be expected, and as may best be seen by reference to the graphs.

ACKNOWLEDGMENTS

During the course of making the large number of observations set forth in this paper, which extended over a period of several years, the authors were from time to time assisted by Mr. C. H. Harper, Research Laboratory, Queens University, Kingston, Ont., now Professor, Moosejaw College, Saskatchewan; and Mr. Walter L. Savell, Research Laboratory, Queens University, Kingston, Ontario, now Metals Department, Deloro Mining and Reduction Company, Deloro, Ontario.

The authors wish, hereby, to acknowledge their indebtedness to these gentlemen, and as well to Mr. R. C. Wilcox, Research Laboratory, Queens University, Kingston, Ontario, now analyst, The Exolon Company, Thorold, Ontario, by whom most of the analyses reported in this paper were made.

CANADA
DEPARTMENT OF MINES
HON. P. E. BLONDIN, MINISTER; R. G. MCCONNELL, DEPUTY MINISTER

MINES BRANCH
EUGENE HAANEL, PH.D., DIRECTOR

REPORTS AND MAPS
PUBLISHED BY THE
MINES BRANCH

REPORTS

†1. Mining conditions in the Klondike, Yukon. Report on—by Eugene Haanel, Ph.D., 1902.

†2. Great landslide at Frank, Alta. Report on—by R. G. McConnell, B.A., and R. W. Brock, M.A., 1903.

†3. Investigation of the different electro-thermic processes for the smelting of iron ores and the making of steel, in operation in Europe. Report of Special Commission—by Eugene Haanel, Ph.D., 1904.

5. On the location and examination of magnetic ore deposits by magnetometric measurements—by Eugene Haanel, Ph.D., 1904.

†7. Limestones, and the lime industry of Manitoba. Preliminary report on—by J. W. Wells, M.A., 1905.

†8. Clays and shales of Manitoba: their industrial value. Preliminary report on—by J. W. Wells, M.A., 1905.

†9. Hydraulic cements (raw materials) in Manitoba; manufacture and uses of. Preliminary report on—by J. W. Wells, M.A., 1905.

†10. Mica: its occurrence, exploitation, and uses—by Fritz Cirkel, M.E., 1905. (See No. 118.)

†11. Asbestos: its occurrence, exploitation, and uses—by Fritz Cirkel, M.E., 1905. (See No. 69.)

†12. Zinc resources of British Columbia and the conditions affecting their exploitation. Report of the Commission appointed to investigate —by W. R. Ingalls, M.E., 1905.

†16. Experiments made at Sault Ste. Marie, under government auspices in the smelting of Canadian iron ore by the electro-thermic process. Final report on—by Eugene Haanel, Ph.D., 1907.

† Publications marked thus † are out of print.

†17. Mines of the silver-cobalt ores of the Cobalt district: their present and prospective output. Report on—by Eugene Haanel, Ph.D., 1907.

†18. Graphite: its properties, occurrences, refining, and uses—by Fritz Cirkel, M.E., 1907.

†19. Peat and lignite: their manufacture and uses in Europe—by Erik Nystrom, M.E., 1908.

†20. Iron ore deposits of Nova Scotia. Report on (Part I)—by J. E. Woodman, D.Sc.

†21. Summary report of Mines Branch, 1907-8.

†22. Iron ore deposits of Thunder Bay and Rainy River districts. Report on—by F. Hille, M.E.

†23. Iron ore deposits along the Ottawa (Quebec side) and Gatineau rivers. Report on—by Fritz Cirkel, M.E.

24. General report on the mining and metallurgical industries of Canada, 1907-8.

†25. The tungsten ores of Canada. Report on—by T. L. Walker, Ph.D.

26. The mineral production of Canada, 1906. Annual report on—by John McLeish, B.A.

†27. The mineral production of Canada, 1907. Preliminary report on—by John McLeish, B.A.

†27a. The mineral production of Canada, 1908. Preliminary report on—by John McLeish, B.A.

†28. Summary report of Mines Branch, 1908.

29. Chrome iron ore deposits of the Eastern Townships. Monograph on—by Fritz Cirkel. (Supplementary section: Experiments with chromite at McGill University—by J. B. Porter, E.M., D.Sc.)

30. Investigation of the peat bogs and peat fuel industry of Canada, 1908, Bulletin No. 1—by Erik Nystrom, M.E., and A. Anrep, Peat Expert.

32. Investigation of electric shaft furnace, Sweden. Report on—by Eugene Haanel, Ph.D.

47. Iron ore deposits of Vancouver and Texada islands. Report on—by Einar Lindeman, M.E.

†55. The bituminous, or oil-shales of New Brunswick and Nova Scotia; also on the oil-shale industry of Scotland. Report on—by R. W. Ells, LL.D.

† Publications marked thus † are out of print.

58. The mineral production of Canada 1907 and 1908. Annual report on—by John McLeish, B.A.

   NOTE.—*The following parts were separately printed and issued in advance of the Annual Report for 1907-08.*

   †31. Production of cement in Canada 190 .

   †42. Production of iron and steel in Canada during the calendar years 1907 and 1908.

   43. Production of chromite in Canada during the calendar years 1907 and 1908.

   44. Production of asbestos in Canada during the calendar years 1907 and 1908.

   †45. Production of coal, coke, and peat in Canada during the calendar years 1907 and 1908.

   46. Production of natural gas and petroleum in Canada during the calendar years 1907 and 1908.

59. Chemical analyses of special economic importance made in the laboratories at the Department of Mines, 1906-07-08. Report on—by F. G. Wait, M.A., F.C.S. (With Appendix on the commercial methods and apparatus for the analyses of oil-shales—by H. A. Leverin, Ch.E.)

   Schedule of charges for chemical analyses and assays.

†62. Mineral production of Canada, 1909. Preliminary report on—by John McLeish, B.A.

63. Summary report of Mines Branch, 1909.

67. Iron deposits of the Bristol mine, Pontiac county, Quebec. Bulletin No. 2—by Einar Lindeman, M.E., and Geo. C. Mackenzie, B.Sc.

†68. Recent advances in the construction of electric furnaces for the production of pig iron, steel, and zinc. Bulletin No. 3—by Eugene Haanel, Ph.D.

69. Chrysotile-asbestos: its occurrence, exploitation, milling, and uses. Report on—by Fritz Cirkel, M.E. (Second edition, enlarged.)

†71. Investigation of the peat bogs and peat industry of Canada, 1909-10; to which is appended Mr. Alf. Larson's paper on Dr. M. Ekenberg's wet-carbonizing process: from Teknisk Tidskrift, No. 12, December 26, 1908—translation by Mr. A. Anrep, Jr.; also a translation of Lieut. Ekelund's pamphlet entitled "A solution of the peat problem," 1909, describing the Ekelund process for the manufacture of peat powder, by Harold A. Leverin, Ch.E. Bulletin No. 4—by A. Anrep. (Second edition, enlarged.)

† Publications marked thus † are out of print.

82. Magnetic concentration experiments. Bulletin No. 5—by Geo. C. Mackenzie, B.Sc.

83. **An** investigation of the coals of Canada with reference to their economic qualities: as conducted at McGill University under the authority of the Dominion Government. Report on—by J. B. Porter, E.M., D.Sc., R. J. Durley, Ma.E., and others.
   Vol. I—Coal washing and coking tests.
   Vol. II—Boiler and gas producer tests.
   †Vol. III—
      Appendix I
         Coal washing tests and diagrams.
   †Vol. IV—
      Appendix II
         Boiler tests and diagrams.
   †Vol. V—
      Appendix III
         Producer tests and diagrams.
   †Vol. VI—
      Appendix IV
         Coking tests.
      Appendix V
         Chemical tests.

†84. Gypsum deposits of the Maritime provinces of Canada—including the Magdalen islands. Report on—by W. F. Jennison, M.E. (See No. 245.)

88. The mineral production of Canada, 1909. Annual report on—by John McLeish, B.A.

> NOTE.—*The following parts were separately printed and issued in advance of the Annual Report for 1909.*

   †79. Production of iron and steel in Canada during the calendar year 1909.
   †80. Production of coal and coke in Canada during the calendar year 1909.
   85. Production of cement, lime, clay products, stone, and other structural materials during the calendar year 1909.

89. Proceedings of conference on explosives. (Fourth edition).

90. Reprint of presidential address delivered before the American Peat Society at Ottawa, July 25, 1910. By Eugene Haanel, Ph.D.

92. Investigation of the explosives industry in the Dominion of Canada, 1910. Report on—by Capt. Arthur Desborough. (Fourth edition).

†93. Molybdenum ores of Canada. Report on—by Professor T. L. Walker Ph.D.

100. The building and ornamental stones of Canada: Building and ornamental stones of Ontario. Report on—by Professor W. A. Parks, Ph.D.

† Publications marked thus † are out of print.

102. Mineral production of Canada, 1910. Preliminary report on—by John McLeish, B.A.

†103. Summary report of Mines Branch, 1910.

104. Catalogue of publications of Mines Branch, from 1902 to 1911; containing tables of contents and lists of maps, etc.

105. Austin Brook iron-bearing district. Report on—by E. Lindeman, M.E.

110. Western portion of Torbrook iron ore deposits, Annapolis county, N.S. Bulletin No. 7—by Howells Frechette, M.Sc.

111. Diamond drilling at Point Mamainse, Ont. Bulletin No. 6—by A. C. Lane, Ph.D., with introductory by A. W. G. Wilson, Ph.D.

118. Mica: its occurrence, exploitation, and uses. Report on—by Hugh S. de Schmid, M.E.

142. Summary report of Mines Branch, 1911.

143. The mineral production of Canada, 1910. Annual report on—by John McLeish, B.A.

> NOTE.—*The following parts were separately printed and issued in advance of the Annual Report for 1910.*
>
> †114. Production of cement, lime, clay products, stone, and other materials in Canada, 1910.
> †115. Production of iron and steel in Canada during the calendar year 1910.
> †116. Production of coal and coke in Canada during the calendar year 1910.
> †117. General summary of the mineral production of Canada during the calendar year 1910.

145. Magnetic iron sands of Natashkwan, Saguenay county, Que. Report on—by Geo. C. Mackenzie, B.Sc.

†150. The mineral production of Canada, 1911. Preliminary report on—by John McLeish, B.A.

151. Investigation of the peat bogs and peat industry of Canada, 1910-11. Bulletin No. 8—by A. Anrep.

154. The utilization of peat for fuel for the production of power, being a record of experiments conducted at the Fuel Testing Station, Ottawa, 1910-11. Report on—by B. F. Haanel, B.Sc.

167. Pyrites in Canada: its occurrence, exploitation, dressing and uses. Report on—by A. W. G. Wilson, Ph.D.

170. The nickel industry: with special reference to the Sudbury region, Ont. Report on—by Professor A. P. Coleman, Ph.D.

† Publications marked thus † are out of print.

184. Magnetite occurrences along the Central Ontario railway. Report on—by E. Lindeman, M.E.

201 The mineral production of Canada during the calendar year 1911 Annual report on—by John McLeish, B.A.

NOTE.—*The following parts were separately printed and issued in advance of the Annual Report for 1911.*

    181. Production of cement, lime, clay products, stone, and other structural materials in Canada during the calendar year 1911. Bulletin on—by John McLeish, B.A.

    †182. Production of iron and steel in Canada during the calendar year 1911. Bulletin on—by John McLeish, B.A.

    183. General summary of the mineral production in Canada during the calendar year 1911. Bulletin on—by John McLeish, B.A.

    †199. Production of copper, gold, lead, nickel, silver, zinc, and other metals of Canada, during the calendar year 1911. Bulletin on—by C. T. Cartwright, B.Sc.

    †200. The production of coal and coke in Canada during the calendar year 1911. Bulletin on—by John McLeish, B.A.

203. Building stones of Canada—Vol. II: Building and ornamental stones of the Maritime Provinces. Report on—by W. A. Parks, Ph.D.

209. The copper smelting industry of Canada. Report on—by A. W. G. Wilson, Ph.D.

216. Mineral production of Canada, 1912. Preliminary report on—by John McLeish, B.A.

222. Lode mining in Yukon: an investigation of the quartz deposits of the Klondike division. Report on—by T. A. MacLean, B.Sc.

224. Summary report of the Mines Branch, 1912.

227. Sections of the Sydney coal fields—by J. G. S. Hudson, M.E.

†229. Summary report of the petroleum and natural gas resources of Canada, 1912—by F. G. Clapp, A.M. (See No. 224).

230. Economic minerals and mining industries of Canada.

245. Gypsum in Canada: its occurrence, exploitation, and technology. Report on—by L. H. Cole, B.Sc.

254. Calabogie iron-bearing district. Report on—by E. Lindeman, M.E.

259. Preparation of metallic cobalt by reduction of the oxide. Report on—by H. T. Kalmus, B.Sc., Ph.D.

† Publications marked thus † are out of print.

262. The mineral production of Canada during the calendar year 1912. Annual report on—by John McLeish, B.A.

> NOTE.—*The following parts were separately printed and issued in advance of the Annual Report for 1912.*
>
> 238. General summary of the mineral production of Canada during the calendar year 1912. Bulletin on—by John McLeish, B.A.
> †247. Production of iron and steel in Canada during the calendar year 1912. Bulletin on—by John McLeish, B.A.
> †256. Production of copper, gold, lead, nickel, silver, zinc, and other metals of Canada, during the calendar year 1912—by C. T. Cartwright, B.Sc.
> 257. Production of cement, lime, clay products, stone, and other structural materials during the calendar year 1912. Report on—by John McLeish, B.A.
> †258. Production of coal and coke in Canada, during the calendar year 1912. Bulletin on—by John McLeish, B.A.

266. Investigation of the peat bogs and peat industry of Canada, 1911 and 1912. Bulletin No. 9—by A. Anrep.

279. Building and ornamental stones of Canada—Vol. III: Building and ornamental stones of Quebec. Report on—by W. A. Parks, Ph.D.

281. The bituminous sands of Northern Alberta. Report on—by S. C. Ells, M.E.

283. Mineral production of Canada, 1913. Preliminary report on—by John McLeish, B.A.

285. Summary report of the Mines Branch, 1913.

291. The petroleum and natural gas resources of Canada. Report on—by F. G. Clapp, A.M., and others:—
Vol. I—Technology and exploitation.
Vol. II—Occurrence of petroleum and natural gas in Canada.
Also separates of Vol. II, as follows:—
Part 1, Eastern Canada.
Part 2, Western Canada.

299. Peat, lignite, and coal: their value as fuels for the production of gas and power in the by-product recovery producer. Report on—by B. F. Haanel, B.Sc.

303. Moose Mountain iron-bearing district. Report on—by E. Lindeman, M.E.

305. The non-metallic minerals used in the Canadian manufacturing industries. Report on—by Howells Fréchette, M.Sc.

309. The physical properties of cobalt, Part II. Report on—by H. T. Kalmus, B.Sc., Ph.D.

† Publications marked thus † are out of print.

320. The mineral production of Canada during the calendar year 1913. Annual report on—by John McLeish, B.A.

> NOTE.—*The following parts were separately printed and issued in advance of the Annual Report for 1913.*
>
> 315. The production of iron and steel during the calendar year 1913. Bulletin on—by John McLeish, B.A.
> †316. The production of coal and coke during the calendar year 1913. Bulletin on—by John McLeish, B.A.
> 317. The production of copper, gold, lead, nickel, silver, zinc, and other metals, during the calendar year 1913. Bulletin on—by C. T. Cartwright, B.Sc.
> 318. The production of cement, lime, clay products, and other structural materials, during the calendar year 1913. Bulletin on—by John McLeish, B.A.
> 319. General summary of the mineral production of Canada during the calendar year 1913. Bulletin on—by John McLeish, B.A.

322. Economic minerals and mining industries of Canada. (Revised Edition).

323. The products and by-products of coal. Report on—by Edgar Stansfield, M.Sc., and F. E. Carter, B.Sc., Dr. Ing.

325. The salt industry of Canada. Report on—by L. H. Cole, B.Sc.

331. The investigation of six samples of Alberta lignites. Report on—by B. F. Haanel, B.Sc., and John Blizard, B.Sc.

333. The mineral production of Canada, 1914. Preliminary report on—by John McLeish, B.A.

334. Electro-plating with cobalt and its alloys. Report on—by H. T. Kalmus, B.Sc., Ph.D.

336. Notes on clay deposits near McMurray, Alberta. Bulletin No. 10—by S. C. Ells, B.A., B.Sc.

338. Coals of Canada: Vol. VII. Weathering of coal. Report on—by J. B. Porter, E.M., Ph.D., D.Sc.

344. Electro-thermic smelting of iron ores in Sweden. Report on—by Alfred Stansfield, D. Sc., A.R.S.M., F.R.S.C.

346. Summary report of the Mines Branch for 1914.

351. Investigation of the peat bogs and the peat industry of Canada, 1913-1914. Bulletin No. 11—by A. Anrep.

384. The Mineral production of Canada during the calendar year 1914. Annual Report on—by John McLeish, B.A.

> NOTE.—*The following parts were separately printed and issued in advance of the Annual Report for 1914.*
>
> 348. Production of coal and coke in Canada during the calendar year, 1914. Bulletin on—by J. McLeish, B.A.

349. Production of iron and steel in Canada during the calendar year, 1914. Bulletin on—by J. McLeish, B.A.
350. Production of copper, gold, lead, nickel, silver, zinc, and other metals, during the calendar year, 1914. Bulletin on—by J. McLeish, B.A.
383. The production of cement, lime, clay products, stone and other structural materials, during the calendar year 1914. Bulletin on—by John McLeish, B.A.
385. Investigation of a reported discovery of phosphate at Banff, Alberta. Bulletin No. 12—by H. S. de Schmid, M.E., 1915.
388. The building and ornamental stones of Canada—Vol. IV: building and ornamental stones of the western provinces. Report on—by W. A. Parks, Ph.D.
401. Feldspar in Canada. Report on—by H. S. de Schmid, M.E.
406. Description of the laboratories of the Mines Branch of the Department of Mines, 1916. Bulletin No. 13.
408. Mineral production of Canada, 1915. Preliminary report on—by John McLeish, B.A.
411. Cobalt alloys with non-corrosive properties. Report on—by H. T. Kalmus, B.Sc., Ph.D.
413. Magnetic properties of cobalt and of Fe_2Co. Report on—by H. T. Kalmus, B.Sc., Ph.D.
419. Production of iron and steel in Canada during the calendar year, 1915. Bulletin on—by J. McLeish, B.A.
421. Summary report of the Mines Branch for 1915.
424. General summary of the mineral production of Canada during the calendar year, 1915. Bulletin on—by John McLeish, B.A.

The Division of Mineral Resources and Statistics has prepared the following lists of mine, smelter, and quarry operators: Metal mines and smelters, General list of mines (except coal and metal mines), Coal mines, Stone quarry operators, Manufacturers of clay products and of cement, Manufacturers of lime, and Operators of sand and gravel deposits. Copies of the lists may be obtained on application.

IN THE PRESS

420. Production of coal and coke in Canada during the calendar year, 1915. Bulletin on—by J. McLeish, B.A.
423. The production of cement, lime, clay products, stone and other structural materials in Canada, during the calendar year, 1915. Bulletin on—by John McLeish, B.A.
425. The production of copper, gold, lead, nickel, silver, zinc, and other metals in Canada, during the calendar year, 1915. Bulletin on—by John McLeish, B.A.
426. The mineral production of Canada during the calendar year, 1915. ial report on—by John McLeish, B.A.
428. The production of spelter in Canada, 1915. Report on—by Dr. A. . Wilson.
430. The coal-fields and coal industry of eastern Canada. Report on—by Francis W. Gray.

FRENCH TRANSLATIONS

971. (26a) Rapport annuel sur les industries minérales du Canada, pour l'année 1905.

†4. Rapport de la Commission nominée pour étudier les divers procédés électro-thermiques pour la réduction des minerais de fer et la fabrication de l'acier employés en Europe—by Eugene Haanel, Ph.D. (French Edition), 1905.

26a. The mineral production of Canada, 1906. Annual report on—by John McLeish, B.A.

†28a. Summary report of Mines Branch, 1908.

56. Bituminous or oil-shales of New Brunswick and Nova Scotia; also on the oil-shale industry of Scotland. Report on—by R. W. Ells, LL.D.

81. Chrysotile-asbestos, its occurrence, exploitation, milling, and uess Report on—by Fritz Cirkel, M.E.

100a. The building and ornamental stones of Canada: Building and ornamental stones of Ontario. Report on—by W. A. Parks, Ph.D.

149. Magnetic iron sands of Natashkwan, Saguenay county, Que. Report on—by Geo. C. Mackenzie, B.Sc.

155. The utilization of peat fuel for the production of power, being a record of experiments conducted at the Fuel Testing Station, Ottawa, 1910-11. Report on—by B. F. Haanel, B.Sc.

†156. The tungsten ores of Canada. Report on—by T. L. Walker, Ph.D.

169. Pyrites in Canada: its occurrences, exploitation, dressing, and uses. Report on—by A. W. G. Wilson, Ph.D.

179. The nickel industry: with special reference to the Sudbury region, Ont. Report on—by Professor A. P. Coleman, Ph.D.

180. Investigation of the peat bogs, and peat industry of Canada, 1910-11. Bulletin No. 8—by A. Anrep.

195. Magnetite occurrences along the Central Ontario railway. Report on —by E. Lindeman, M.E.

†196. Investigation of the peat bogs and peat industry of Canada, 1909-10, to which is appended Mr. Alf. Larson's paper on Dr. M. Ekenburg's wet-carbonizing process: from Teknisk Tidskrift, No. 12, December 26, 1908—translation by Mr. A. Anrep; also a translation of Lieut. Ekelun's, pamphlet entitled "A solution of the peat problem," 1909, describing the Ekelund process for the manufacture of peat powder, by Harold A. Leverin, Ch.E. Bulletin No. 4—by A. Anrep. (Second Edition, enlarged.)

197. Molybdenum ores of Canada. Report on—by T. L. Walker, Ph.D.

† Publications marked thus † are out of print.

†198. Peat and lignite: their manufacture and uses in Europe. Report on—by Erik Nystrom, M.E., 1908.

202. Graphite: its properties, occurrences, refining, and uses. Report on—by Fritz Cirkel, M.E., 1907.

204. Building stones of Canada—Vol. II: Building and ornamental stones of the Maritime Provinces. Report on—by W. A. Parks, Ph.D.

219. Austin Brook iron-bearing district. Report on—by E. Lindeman, M.E.

223. Lode Mining in the Yukon: an investigation of quartz deposits in the Klondike division. Report on—by T. A. MacLean, B.Sc.

224a. Mines Branch Summary report for 1912.

†226. Chrome iron ore deposits of the Eastern Townships. Monograph on—by Fritz Cirkel, M.E. (Supplementary section: Experiments with chromite at McGill University—by J. B. Porter, E.M., D.Sc.).

231. Economic minerals and mining industries of Canada.

233. Gypsum deposits of the Maritime Provinces of Canada—including the Magdalen islands. Report on—by W. F. Jennison, M.E.

246. Gypsum in Canada: its occurrence, exploitation, and technology. Report on—by L. H. Cole, B.Sc.

260. The preparation of metallic cobalt by reduction of the oxide. Report on—by H. T. Kalmus, B.Sc., Ph.D.

263. Recent advances in the construction of electric furnaces for the production of pig iron, steel, and zinc. Bulletin No. 3—by Eugene Haanel, Ph.D.

†264. Mica: its occurrence, exploitation, and uses. Report on—by Hugh S. de Schmid, M.E.

265. Annual mineral production of Canada, 1911. Report on—by John McLeish, B.A.

280. The building and ornamental stones of Canada, Vol. III; Province of Quebec. Report on—by Professor W. A. Parks, Ph.D.

282. The bituminous sands of Northern Alberta. Report on—by S. C. Ells, M.E.

286. Summary Report of Mines Branch, 1913.

287. Production of iron and steel in Canada during the calendar year 1912. Bulletin on—by John McLeish, B.A.

288. Production of coal and coke in Canada, during the calendar year 1912. Bulletin on—by John McLeish, B.A.

† Publications marked thus † are out of print.

289. Production of cement, lime, clay products, stone, and other structural materials during the calendar year 1912. Bulletin on—by John McLeish, B.A.

290. Production of copper, gold, lead, nickel, silver, zinc, and other metals of Canada during the calendar year 1912. Bulletin on—by C. T. Cartwright, B.Sc.

307. Cataologue of French publications of the Mines Branch and of the Geological Survey, up to July, 1914.

308. An investigation of the coals of Canada with reference to their economic qualities: as conducted at McGill University under the authority of the Dominion Government. Report on—by J. B. Porter, E.M., D.Sc., R. J. Durley, Ma.E., and others.
  Vol. I—Coal washing and coking tests.
  Vol. II—Boiler and gas producer tests.
  Vol. III—
    Appendix I
      Coal washing tests and diagrams.
  Vol. IV—
    Appendix II
      Boiler tests and diagrams.

314. Iron ore deposits, Bristol mine, Pontiac county, Quebec, Report on—by E. Lindeman, M.E.

321. Annual mineral production of Canada, during the calendar year 1913. Report on—by J. McLeish, B.A.

415. Annual mineral production of Canada during the calendar year, 1914. Report on—by J. McLeish, B.A.

IN THE PRESS

292. The petroleum and natural gas resources of Canada. Report on—by F. G. Clapp, A.M., and others.
  Vol. I.—Technology and exploitation.

306. The non-metallic minerals used in the Canadian manufacturing industries Report on—by Howells Fréchette, M.Sc.

310. The physical properties of the metal cobalt, Part II. Report on—by H. T. Kalmus, B.Sc., Ph.D.

324. Products and by-products of coal. Report on—by Edgar Stansfield M.Sc., and F. E. Carter, B.Sc., Dr. Ing.

326. The salt industry of Canada. Report on—by L. H. Cole, B.Sc.

MAPS

†6. Magnetometric survey, vertical intensity: Calabogie mine, Bagot township, Renfrew county, Ontario—by E. Nystrom, 1904. Scale 60 feet to 1 inch. Summary report 1905. (See Map No. 249.)

†13. Magnetometric survey of the Belmont iron mines, Belmont township, Peterborough county, Ontario—by B. F. Haanel, 1905. Scale 60 feet to 1 inch. Summary report, 1906. (See Map No. 186).

†14. Magnetometric survey of the Wilbur mine, Lavant township, Lanark county, Ontario—by B. F. Haanel, 1905. Scale 60 feet to 1 inch. Summary report, 1906.

†33. Magnetometric survey, vertical intensity; lot 1, concession VI, Mayo township, Hastings county, Ontario—by Howells Fréchette, 1909. Scale 60 feet to 1 inch. (See Maps Nos. 191 and 191A.)

†34. Magnetometric survey, vertical intensity; lots 2 and 3, concession VI, Mayo township, Hastings county, Ontario—by Howells Fréchette, 1909. Scale 60 feet to 1 inch. (See Maps Nos. 191 and 191A.)

†35. Magnetometric survey, vertical intensity: lots 10, 11, and 12 concession IX, and lots 11 and 12, concession VIII, Mayo township, Hastings county, Ontario—by Howells Fréchette, 1909. Scale 60 feet to 1 inch. (See Maps Nos. 191 and 191A.)

*36. Survey of Mer Bleue peat bog, Gloucester township, Carleton county, and Cumberland township, Russell county, Ontario—by Erik Nystrom, and A. Anrep. (Accompanying report No. 30.)

*37. Survey of Alfred peat bog. Alfred and Caledonia townships, Prescott county, Ontario—by Erik Nystrom and A. Anrep. (Accompanying report No. 30.)

*38. Survey of Welland peat bog, Wainfleet and Humberstone townships, Welland county, Ontario—by Erik Nystrom and A. Anrep. (Accompanying report No. 30.)

*39. Survey of Newington peat bog, Osnabruck, Roxborough, and Cornwall townships, Stormont county, Ontario—by Erik Nystrom and A. Anrep. (Accompanying report No. 30.)

*40. Survey of Perth peat bog, Drummond township, Lanark county, Ontario—by Erik Nystrom and A. Anrep. (Accompanying report No. 30.)

†41. Survey of Victoria Road peat bog, Bexley and Carden townships, Victoria county, Ontario—Erik Nystrom and A. Anrep. (Accompanying report No. 30).

*48. Magnetometric survey of Iron Crown claim at Nimpkish (Klaanch) river, Vancouver island, B.C.—by E. Lindeman. Scale 60 feet to 1 inch. (Accompanying report No. 47).

Note.—1. Maps marked thus * are to be found only in reports.
2. Maps marked thus † have been printed independently of reports, hence can be procured separately by applicants.

xiv

*49. Magnetometric survey of Western Steel Iron claim, at Sechart, Vancouver island, B.C.—By E. Lindeman. Scale 60 feet to 1 inch. (Accompanying report No. 47.)

*53. Iron ore occurrences, Ottawa and Pontiac counties, Quebec, 1908—by J. White and Fritz Cirkel. (Accompanying report No. 23.)

*54. Iron ore occurrences, Argenteuil county, Quebec, 1908—by Fritz Cirkel. (Accompanying report No. 23.) Out of print.

†57. The productive chrome iron ore district of Quebec—by Fritz ...l. (Accompanying report No. 29.)

†60. Magnetometric survey of the Bristol mine, Pontiac county, Quebec— by E. Lindema . Scale 200 feet to 1 inch. (Accompanying report No. 67.)

†61. Topographical map of Bristol mine, Pontiac county, Quebec—by E. Lindeman. Scale 200 feet to 1 inch. (Accompanying report No. 67.)

†64. Index map of Nova Scotia: Gypsum—by W. F. Jennison. ⎫
†65. Index map of New Brunswick: Gypsum—by W. F. Jenni- ⎬ (Accompanying report No. 84.)
son.   ⎪
†66. Map of Magdalen islands: Gypsum—by W. F. Jennison.... ⎭

†70. Magnetometric survey of Northeast Arm iron range, Lake Timagami, Nipissing district, Ontario—by E. Lindeman. Scale 200 feet to 1 inch. (Accompanying report No 63.)

†72. Brunner peat bog, Ontario—by A. Anrep.   ⎫
†73. Komoka peat bog, Ontario—   "   "   ⎬ (Accompanying report No. 71.)
†74. Brockville peat bog, Ontario— "   "   ⎭

†75. Rondeau peat bog, Ontario—   "   "   ⎫ (Out of
†76. Alfred peat bog, Ontario—   "   "   ⎭ print.)

†77. Alfred peat bog, Ontario, main ditch profile—by A. Anrep.

†78. Map of asbestos region, Province of Quebec, 1910—by Fritz Cirkel. Scale 1 mile to 1 inch. (Accompanying report No. 69.)

†94. Map showing Cobalt, Gowganda, Shiningtree, and Porcupine districts —by L. H. Cole. (Accompanying Summary report, 1910.)

†95. General map of Canada, showing coal fields. (Accompanying report No. 83—by Dr. J. B. Porter.)

†96. General map of coal fields of Nova Scotia and New Brunswick. (Accompanying report No. 83—by Dr. J. B. Porter.)

†97. General map showing coal fields in Alberta, Saskatchewan, and Manitoba. (Accompanying report No. 83—by Dr. J. B. Porter.)

Note.—1. Maps marked thus * are to be found only in reports.
2. Maps marked thus † have been printed independently of reports, hence can be procured separately by applicants.

xv

†98. **General** map of coal fields in British Columbia. (Accompanying report No. 83—by Dr. J. B. Porter.)

†99. **General** map of coal field in Yukon Territory. (Accompanying report No. 83—by Dr. J. B. Porter.)

†106. **Geological** map of Austin Brook iron-bearing district, Bathurst township, Gloucester county, N.B.—by E. Lindeman. Scale 400 feet to 1 inch. (Accompanying report No. 105.)

†107. **Magnetometric** survey, vertical intensity; Austin Brook iron-bearing district—by E. Lindeman. Scale 400 feet to 1 inch. (Accompanying report No. 105.)

†108. **Index** map showing iron-bearing area at Austin Brook—by E. Lindeman. (Accompanying report No. 105.)

*112. **Sketch** plan showing geology of Point Mamainse, Ont.—by Professor A. C. Lane. Scale 4,000 feet to 1 inch. (Accompanying report No. 111.)

†113. **Holland** peat bog Ontario—by A. Anrep. (Accompanying report No. 1

*119–137. **Mica**: township maps, Ontario and Quebec—by Hugh S. de Schmid. (Accompanying report No. 118.)

†138. **Mica**: showing location of principal mines and occurrences in the Quebec mica area—by Hugh S. de Schmid. Scale 3·95 miles to 1 inch. (Accompanying report No. 118.)

†139. **Mica**: showing location of principal mines and occurrences in the Ontario mica area—by Hugh S. de Schmid. Scale 3·95 miles to 1 inch. (Accompanying report No. 118.)

†140. **Mica**: showing distribution of the principal mica occurrences in the Dominion of Canada—by Hugh S. de Schmid. Scale 3·95 miles to 1 inch. (Accompanying report No. 118.)

†141. **Torbrook** iron-bearing district Annapolis county, N.S.—by Howells Fréchette. Scale 400 feet to 1 inch. (Accompanying report No. 110.)

146. **Distribution** of iron ore sands of the iron ore deposits on the north shore of the River and Gulf of St. Lawrence, Canada—by Geo. C. Mackenzie. Scale 100 miles to 1 inch. (Accompanying report No. 145.)

†147. **Magnetic** iron sand deposits in relation to Natashkwan harbour and Great Natashkwan river, Que. (Index Map)—by Geo. C. Mackenzie. Scale 40 chains to 1 inch. (Accompanying report No. 145.)

†148. **Natashkwan** magnetic iron sand deposits, Saguenay county, Que.—by Geo. C. Mackenzie. Scale 1,000 feet to 1 inch. (Accompanying report No. 145.)

Note.—1. Maps marked thus * are to be found only in reports.
2. Maps marked thus † have been printed independently of reports, hence can be procured separately by applicants.

†152. Map showing the location of peat bogs investigated in Ontario—by A. Anrep. (See Map No. 154.)

†153. Map showing the location of peat bogs, as investigated in Manitoba—by A. Anrep.

†157. Lac du Bonnet peat bog, Manitoba—by A. Anrep.

†158. Transmission peat bog, Manitoba— " "

†159. Corduroy peat bog, Manitoba— " "

†160. Boggy Creek peat bog, Manitoba— " "

†161. Rice Lake peat bog, Manitoba— " "

†162. Mud Lake peat bog, Manitoba— " "

†163. Litter peat bog, Manitoba— " "

†164. Julius peat litter bog, Manitoba— " "

†165. Fort Frances peat bog, Ontario— " "

(Accompanying report No. 151.)

*166. Magnetometric map of No. 3 mine, lot 7, concessions V and VI, McKim township, Sudbury district, Ont.—by E. Lindeman. (Accompanying Summary report, 1911.)

†168. Map showing pyrite, mines and prospects in Eastern Canada, and their relation to the United States market—by A. W. G. Wilson. Scale 125 miles to 1 inch. (Accompanying report No. 167.)

†171. Geological map of Sudbury nickel region, Ont.—by Prof. A. P. Coleman. Scale 1 mile to 1 inch. (Accompanying report No. 170.)

†172. Geological map of Victoria mine—by Prof. A. P. Coleman.

†173. " Crean Hill mine—by Prof. A. P. Coleman.

†174. " Creighton mine—by Prof. A. P. Coleman.

(Accompanying report No. 170.)

†175. " showing contact of norite and Laurentian in vicinity of Creighton mine—by Prof. A. P. Coleman. (Accompanying report No. 170.)

†176. " Copper Cliff offset—by Prof. A. P. Coleman. (Accompanying report No. 170.)

†177. " No. 3 mine—by Prof. A. P. Coleman. (Accompanying report No. 170.)

†178. " showing vicinity of Stobie and No. 3 mines—by Prof. A. P. Coleman. (Accompanying report No. 170.)

Note.—1. Maps marked thus * are to be found only in reports.
2. Maps marked thus † have been printed independently of reports, hence can be procured separately by applicants.

xvii

†185. Magnetometric survey, vertical intensity: Blairton iron mine, Belmont township, Peterborough county, Ontario—by E. Lindeman, 1911. Scale 200 feet to 1 inch. (Accompanying report No. 184.)

†185a. Geological map, Blairton iron mine, Belmont township, Peterborough county, Ontario—by E. Lindeman, 1911. Scale 200 feet to 1 inch. (Accompanying report No. 184.)

†186. Magnetometric survey, Belmont iron mine, Belmont township, Peterborough county, Ontario—by E. Lindeman, 1911. Scale 200 feet to 1 inch. (Accompanying report No. 184.)

†186a. Geological map, Belmont iron mine, Belmont township, Peterborough county, Ontario—by E. Lindeman, 1911. Scale 200 feet to 1 inch. (Accompanying report No. 184.)

†187. Magnetometric survey, vertical intensity: St. Charles mine, Tudor township, Hastings county, Ontario—by E. Lindeman, 1911. Scale 200 feet to 1 inch. (Accompanying report No. 184.)

†187a. Geological map, St. Charles mine, Tudor township, Hastings county, Ontario—by E. Lindeman, 1911. Scale 200 feet to 1 inch. (Accompanying report No. 184.)

†188. Magnetometric survey, vertical intensity: Baker mine, Tudor township, Hastings county, Ontario—by E. Lindeman, 1911. Scale 200 feet to 1 inch. (Accompanying report No. 184.)

†188a. Geological map, Baker mine, Tudor township, Hastings county, Ontario—by E. Lindeman, 1911. Scale 200 feet to 1 inch. (Accompanying report No. 184.)

†189. Magnetometric survey, vertical intensity: Ridge iron ore deposits, Wollaston township, Hastings county, Ontario—by E. Lindeman, 1911. Scale 200 feet to 1 inch. (Accompanying report No. 184.)

†190. Magnetometric survey, vertical intensity: Coehill and Jenkins mines, Wollaston township, Hastings county, Ontario—by E. Lindeman, 1911. Scale 200 feet to 1 inch. (Accompanying report No. 184.)

†190a. Geological map, Coehill and Jenkins mines, Wollaston township, Hastings county, Ontario—by E. Lindeman, 1911. Scale 200 feet to 1 inch. (Accompanying report No. 184.)

†191. Magnetometric survey, vertical intensity: Bessemer iron ore deposits, Mayo township, Hastings county, Ontario—by E. Lindeman, 1911. Scale 200 feet to 1 inch. (Accompanying report No. 184.)

†191a. Geological map, Bessemer iron ore deposits, Mayo township, Hastings county, Ontario—by E. Lindeman, 1911. Scale 200 feet to 1 inch. (Accompanying report No. 184.)

†192. Magnetometric survey, vertical intensity: Rankin, Childs, and Stevens mines, Mayo township, Hastings county, Ontario—by E. Lindeman, 1911. Scale 200 feet to 1 inch. (Accompanying report No. 184.)

Note.—1. Maps marked thus * are to be found only in reports.
2. Maps marked thus † have been printed independently of reports, hence can be procured separately by applicants.

†192a. Geological map, Rankin, Childs, and Stevens mines, Mayo township, Hastings county, Ontario—by E. Lindeman, 1911. Scale 200 feet to 1 inch. (Accompanying report No. 184.)

†193. Magnetometric survey, vertical intensity: Kennedy property, Carlow township, Hastings county, Ontario—by E. Lindeman, 1911. Scale 200 feet to 1 inch. (Accompanying report No. 184.)

†193a. Geological map, Kennedy property, Carlow township, Hastings county Ontario—by E. Lindeman, 1911. Scale 200 feet to 1 inch. (Accompanying report No. 184.)

†194. Magnetometric survey, vertical intensity: Bow Lake iron ore occurrences, Faraday township, Hastings county, Ontario—by E. Lindeman, 1911. Scale 200 feet to 1 inch. (Accompanying report No. 184.)

†204. Index map, magnetite occurrences along the Central Ontario railway—by E. Lindeman, 1911. (Accompanying report No. 184.)

†205. Magnetometric map, Moose Mountain iron-bearing district, Sudbury district, Ontario: Deposits Nos. 1, 2, 3, 4, 5, 6, and 7—by E. Lindeman, 1911. (Accompanying report No. 303.)

†205a. Geological map, Moose Mountain iron-bearing district, Sudbury district, Ontario, Deposits Nos. 1, 2, 3, 4, 5, 6, and 7—by E. Lindeman. (Accompanying report No. 303.)

†206. Magnetometric survey of Moose Mountain iron-bearing district, Sudbury district, Ontario: northern part of deposit No. 2—by E Lindeman, 1912. Scale 200 feet to 1 inch. (Accompanying report No. 303.)

†207. Magnetometric survey of Moose Mountain iron-bearing district, Sudbury district, Ontario: Deposits Nos. 8, 9, and 9A—by E. Lindeman, 1912. Scale 200 feet to 1 inch. (Accompanying report No. 303.)

†208. Magnetometric survey of Moose Mountain iron-bearing district, Sudbury district, Ontario: Deposit No. 10—by E. Lindeman, 1912. Scale 200 feet to 1 inch. (Accompanying report No. 303.)

†208a. Magnetometric survey, Moose Mountain iron-bearing district, Sudbury district, Ontario: eastern portion of Deposit No. 11—by E. Lindeman, 1912. Scale 200 feet to 1 inch. (Accompanying report No. 303.)

†208b. Magnetometric survey, Moose Mountain iron-bearing district, Sudbury district, Ontario: western portion of deposit No. 11—by E. Lindeman, 1912. Scale 200 feet to 1 inch. (Accompanying report No. 303.)

†208c. General geological map, Moose Mountain iron-bearing district, Sudbury district, Ontario—by E. Lindeman, 1912. Scale 800 feet to 1 inch. (Accompanying report No. 303.)

Note.—1. Maps marked thus * are to be found only in reports.
2. Maps marked thus † have been printed independently of reports hence can be procured separately by applicants.

*210. Location of copper smelters in Canada—by A. W. G. Wilson. Scale 197·3 miles to 1 inch. (Accompanying report No. 209.)

†215. Province of Alberta: showing properties from which samples of coal were taken for gas producer tests, Fuel Testing Division, Ottawa. (Accompanying Summary report, 1912.)

†220. Mining districts, Yukon. Scale 35 miles to 1 inch—by T. A. MacLean. (Accompanying report No. 222.)

†221. Dawson mining district, Yukon. Scale 2 miles to 1 inch—by T. A. MacLean. (Accompanying report No. 222.)

*228. Index map of the Sydney coal fields, Cape Breton, N.S. (Accompanying report No. 227.)

†232. Mineral map of Canada. Scale 100 miles to 1 inch. (Accompanying report No. 230.)

†239. Index map of Canada showing gypsum occurrences. (Accompanying report No. 245.)

†240. Map showing Lower Carboniferous formation in which gypsum occurs in the Maritime provinces. Scale 100 miles to 1 inch. (Accompanying report No. 345.)

†241. Map showing relation of gypsum deposits in Northern Ontario to railway lines. Scale 100 miles to 1 inch. (Accompanying report No. 245.)

†242. Map, Grand River gypsum deposits, Ontario. Scale 4 miles to 1 inch. (Accompanying report No. 245.)

†243. Plan of Manitoba Gypsum Co.'s properties. (Accompanying report No. 245.)

†244. Map showing relation of gypsum deposits in British Columbia to railway lines and market. Scale 35 miles to 1 inch. (Accompanying report No. 245.)

†249. Magnetometric survey, Caldwell and Campbell mines, Calabogie district, Renfrew county, Ontario—by E. Lindeman, 1911. Scale 200 feet to 1 inch. (Accompanying report No. 254.)

†250. Magnetometric survey, Black Bay or Williams mine, Calabogie district, Renfrew county, Ontario—by E. Lindeman, 1911. Scale 200 feet to 1 inch. (Accompanying report No. 254.)

†251. Magnetometric survey, Bluff Point iron mine, Calabogie district, Renfrew county, Ontario—by E. Lindeman, 1911. Scale 200 feet to 1 inch. (Accompanying report No. 254.)

†252. Magnetometric survey, Culhane mine, Calabogie district, Renfrew county, Ontario—by E. Lindeman, 1911. Scale 200 feet to 1 inch. (Accompanying report No. 254.)

Note.—1. Maps marked thus * are to be found only in reports.
2. Maps marked thus † have been printed independently of reports, hence can be procured separately by applicants.

†253. Magnetometric survey, Martel or Wilson iron mine, Calabogie district, Renfrew county, Ontario—by E. Lindeman, 1911. Scale 200 feet to 1 inch. (Accompanying report No. 254.)

†261. Magnetometric survey, Northeast Arm iron range, lot 339 E.T.W. Lake Timagami, Nipissing district, Ontario—by E. Nystrom. 1903. Scale 200 feet to 1 inch.

†268. Map of peat bogs investigated in Quebec—by A. Anrep, 1912. (See Report No. 351.)

†269. Large Tea Field peat bog, Quebec " " "

†270. Small Tea Field peat bog, Quebec " " "

†271. Lanoraie peat bog, Quebec " " "

†272. St. Hyacinthe peat bog, Quebec " " "

†273. Rivière du Loup peat bog " " "

†274. Cacouna peat bog " " "

†275. Le Parc t bog, Quebec " " "

†276. St. De. peat bog, Quebec " " "

†277. Rivière Ouelle peat bog, Quebec " " "

†278. Moose Mountain peat bog, Quebec " " "

†284. Map of northern portion of Alberta, showing position of outcrops of bituminous sand. Scale 12½ miles to 1 inch. (Accompanying report No. 281.)

†293. Map of Dominion of Canada, showing the occurrences of oil, gas, and tar sands. Scale 197 miles to 1 inch. (Accompanying report No. 291.)

†294. Reconnaissance map of part of Albert and Westmorland counties, New Brunswick. Scale 1 mile to 1 inch. (Accompanying report No. 291.)

†295. Sketch plan of Gaspé oil fields, Quebec, showing location of wells. Scale 2 miles to 1 inch. (Accompanying report No. 291.)

†296. Map showing gas and oil fields and pipe-lines in southwestern Ontario. Scale 4 miles to 1 inch. (Accompanying report No. 291.)

†297. Geological map of Alberta, Saskatchewan, and Manitoba. Scale 35 miles to 1 inch. (Accompanying report No. 291.)

†298. Map, geology of the forty-ninth parallel, 0·9864 miles to 1 inch. (Ac npanying report No. 291.)

Note.—1. Maps marked thus * are to be found only in reports.
2. Maps marked thus † have been printed independently of reports, hence can be procured separately by applicants.

†302. Map showing location of main gas line, Bow Island, Calgary. Scale 12½ miles to 1 inch. (Accompanying report No. 291.)

†311. Magnetometric map, McPherson mine, Barachois, Cape Breton county, Nova Scotia—by A. H. A. Robinson, 1913. Scale 200 feet to 1 inch.

†312. Magnetometric map, iron ore deposits at Upper Glencoe, Inverness county, Nova Scotia—by E. Lindeman, 1913. Scale 200 feet to 1 inch.

†313. Magnetometric map, iron ore deposits at Grand Mira, Cape Breton county, Nova Scotia—by A. H. A. Robinson, 1913. Scale 200 feet to 1 inch.

†327. Map showing location of Saline Springs and Salt Areas in the Dominion of Canada. (Accompanying Report No. 325.)

†328. Map showing location of Saline Springs in the Maritime Provinces. Scale 100 miles to 1 inch. (Accompanying Report No. 325.)

†329. Map of Ontario-Michigan Salt Basin, showing probable limit of productive area. Scale 25 miles to 1 inch. (Accompanying Report No. 325.)

†330. Map showing location of Saline Springs in Northern Manitoba. Scale 12½ miles to 1 inch. (Accompanying Report No. 325.)

†340. Magnetometric map of Atikokan iron-bearing district, Atikokan Mine and Vicinity. Claims Nos. 10E, 11E, 12E, 24E, 25E, and 26E, Rainy River district, Ontario. By A. H. A. Robinson, 1914. Scale 400 feet to 1 inch.

†340a. Geological map of Atikokan iron-bearing district, Atikokan Mine and Vicinity. Claims Nos. 10E, 11E, 12E, 24E, 25E, and 26E, Rainy River district, Ontario. By A. H. A. Robinson, 1914. Scale 400 feet to 1 inch.

†341. Magnetometric map of Atikokan iron-bearing district, Sheet No. 1, Claims Nos. 400R, 401R, 402R, 112X, and 403R. Rainy River district, Ontario. By E. Lindeman, 1914. Scale 400 feet to 1 inch.

†341a. Geological map of Atikokan iron-bearing district. Sheet No. 1. Claims Nos. 400R, 401R, 402R, 112X, and 403R, Rainy River district, Ontario. By E. Lindeman, 1914. Scale 400 feet to 1 inch.

†342. Magnetometric map of Atikokan iron-bearing district. Sheet No. 2. Claims Nos. 403R, 404R, 138X, 139X, and 140X, Rainy River district, Ontario. By E. Lindeman, 1914. Scale 400 feet to 1 inch.

†342a. Geological map of Atikokan iron-bearing district. Sheet No. 2. Claims Nos. 403R, 404R, 138X, 139X, and 140X, Rainy River district, Ontario. By E. Lindeman, 1914. Scale 400 feet to 1 inch.

† Maps marked thus † have been printed independently of reports, hence can be procured separately by applicants.

†343. Magnetometric map of Atikokan iron-bearing district. Mile Post No. 140, Canadian Northern railway, Rainy River district, Ontario. By E. Lindeman, 1914. Scale 400 feet to 1 inch.

†343a. Geological map, Atikokan iron-bearing district. Mile Post No. 140, Canadian Northern railway, Rainy River district, Ontario. By E. Lindeman, 1914. Scale 400 feet to 1 inch.

†354. Index Map, showing location of peat bogs investigated in Ontario— by A. Anrep, 1913-14.

†355. Richmond peat bog, Carleton county, Ontario— " "

†356. Luther peat bog, Wellington and Dufferin counties, Ontario— " "

†357. Amaranth peat bog, Dufferin county, Ontario— " "

†358. Cargill peat bog, Bruce county, Ontario— " "

†359. Westover peat bog, Wentworth county, Ontario— " "

†360. Marsh Hill peat bog, Ontario county, Ontario— " "

†361. Sunderland peat bog, Ontario county, Ontario— " "

†362. Manilla peat bog, Victoria county, Ontario— " "

†363. Stoco peat bog, Hastings county, Ontario— " "

†364. Clareview peat bog, Lennox and Addington counties, Ontario— " "

†365. Index Map, showing location of peat bogs investigated in Quebec— " "

†366. L'Assomption peat bog, L'Assomption county, Quebec— " "

†367. St. Isidore peat bog, La Prairie county, Quebec— " "

†368. Holton peat bog, Chateauguay county, Quebec— " "

†369. Index Map, showing location of peat bogs investigated in Nova Scotia and Prince Edward Island— " "

†370. Black Marsh peat bog, Prince county, Prince Edward Island— " "

†371. Portage peat bog, Prince county, Prince Edward Island— " "

†372. Miscouche peat bog, Prince county, Prince Edward Island— " "

†373. Muddy Creek peat bog, Prince county, Prince Edward Island— " "

†374. The Black Banks peat bog, Prince county, Prince Edward Island— " "

†Maps marked thus † have been printed independently of reports, hence can be procu separately by applicants.

†375. Mermaid peat bog, Queens county, Prince
Edward Island.......................by A. Anrep, 1913-14

†376. Caribou peat bog, Kings county, Prince Edward
Island—   " "

†377. Cherryfield peat bog, Lunenburg County, Nova
Scotia—   " "

†378. Tusket peat bog, Yarmouth county, Nova
Scotia—   " "

†379. Makoke peat bog, Yarmouth county, Nova
Scotia—   " "

†380. Heath peat bog, Yarmouth county, Nova
Scotia—   " "

†381. Port Clyde peat bog, Shelburne county, Nova
Scotia—   " "

†382. Latour peat bog, Shelburne county, Nova
Scotia—   " "

†383. Clyde peat bog, Shelburne county, Nova
Scotia—   " "

†387. Geological map Banff district, Alberta, showing location of phosphate beds—by Hugh S. de Schmid, 1915. (Accompanying report No. 385.)

†390. Christina river map showing outcrops of bituminous sand along Christina valley; contour intervals of 20 feet—by S. C. Ells, 1915. Scale 1,000 feet to 1 inch.

†391. Clearwater river map, showing outcrops of bituminous sand along Clearwater valley; contour intervals of 20 feet—by S. C. Ells, 1915. Scale 1,000 feet to 1 inch.

†392. Hangingstone-Horse rivers, showing outcrops of bituminous sand along Hangingstone and Horse River valleys; contour intervals of 20 feet—by S. C. Ells, 1915. Scale 1,000 feet to 1 inch.

†393. Steepbank river, showing outcrops of bituminous sand along Steepbank valley; contour intervals of 20 feet—by S. C. Ells, 1915. Scale 1,000 feet to 1 inch.

†394. McKay river, 3 sheets, showing outcrops of bituminous sand along McKay valley; contour intervals of 20 feet—by S. C. Ells, 1915. Scale 1,000 feet to 1 inch.

†395. Moose river, showing outcrops of bituminous sand along Moose valley; contour intervals of 20 feet—by S. C. Ells, 1915. Scale 1,000 feet to 1 inch.

398. Ontario phosphate area—by Hugh S. deSchmid.

399. Quebec phosphate area—   " " " "

403. Ontario feldspar area—   " " " "

404. Quebec feldspar area—   " " " "

† Maps marked thus † have been printed independently of reports, hence can be procured separately by applicants.

405. Magnetometric map, Orton mine and vicinity, Hastings county, Ontario—by A. H. A. Robinson, 1915.

409. Magnetometric map, Kaministikwia, Thunder Bay district, Ontario—by A. H. A. Robinson, 1914-15.

410. Geological map, Kaministikwia, Thunder Bay district, Ontario—by A. H. A. Robinson, 1914-15.

416. Magnetometric map, Matawin Iron Range, Thunder Bay district, Ontario—by A. H. A. Robinson, 1914-15.

Address all communications to—

DIRECTOR MINES BRANCH,
DEPARTMENT OF MINES,
SUSSEX STREET, OTTAWA.

Note.—1. Maps marked thus * are to be found only in reports.
2. Maps marked thus † have been printed independently of reports, hence can be procured separately by applicants.

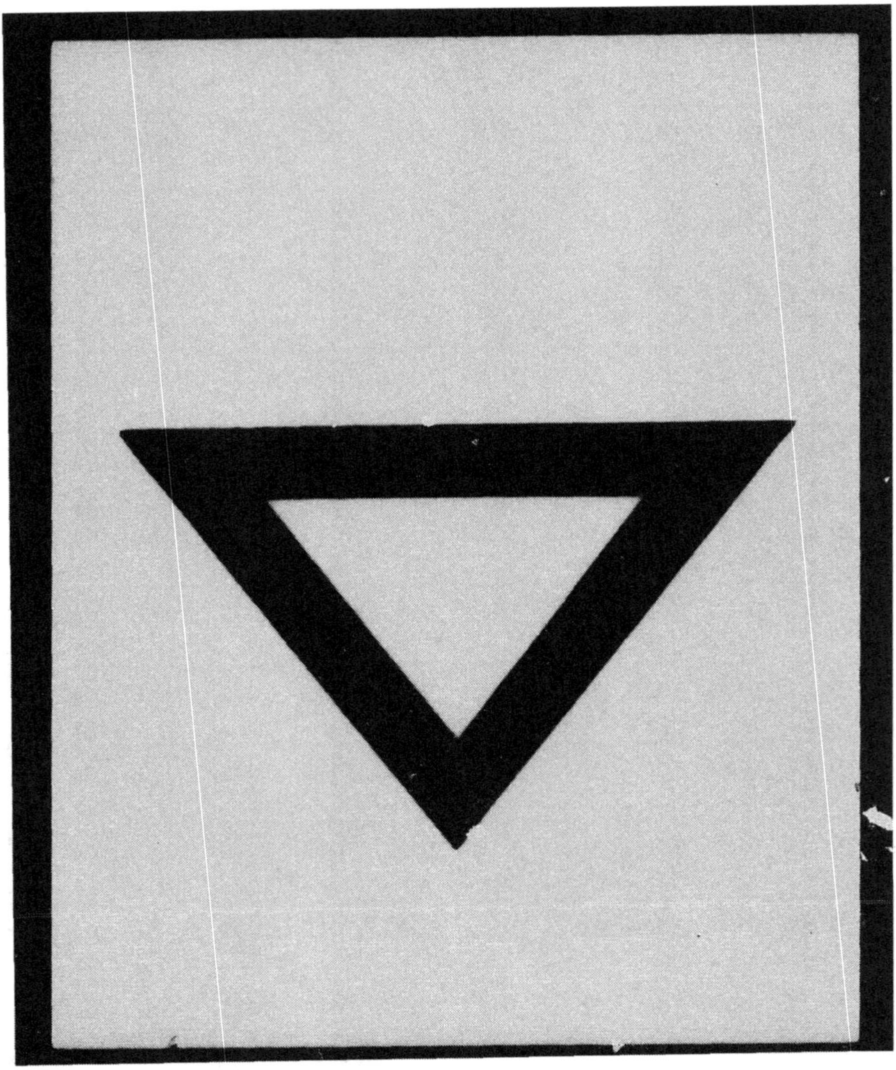